Saline Water Distillation Processes

Saline water distillation processes

Andrew Porteous
BSc, MEng, DEng, CEng, MASME, MIChemE, MIMechE

Reader in Engineering Mechanics
at the Open University

Longman

Longman
1724-1974

Longman Group Limited
London
Associated companies, branches and representatives
throughout the world

First published 1975

ISBN 0 582 46227 4

Library of Congress Catalogue Card No. 73-93117

Set in IBM Press Roman 10/12pt
and printed in Great Britain
by Lowe & Brydone (Printers) Ltd,
Thetford, Norfolk

To Margaret and Neil

Contents

Notation	ix
Preface	xii
Acknowledgements	xiii
Chapter 1 A very general introduction	1
Characteristics of life	1
The first and second laws of thermodynamics	2
Energy resources	3
The hydrological cycle	4
Water supply and demand	5
Water quality standards	5
Desalination processes and their applicability	8
Process summary	10
Electrodialysis	10
Reverse osmosis	10
Freezing	11
Distillation	12
Chapter 2 Fundamentals	14
Introduction	14
Heat transfer	15
Heat transfer surface and performance ratio	18
Boiling point elevation	20
Pressure-drop losses	22
Hydrostatic head effects	22
The second law of thermodynamics	23
Flash range	24
References	24
Chapter 3 The control of scale and corrosion	25
Introduction	25
Composition and properties of typical raw feed-water	25

Alkaline scales 27
Methods of scale control 30
Calcium sulphate scaling 34
Feed treatment selection 38
Corrosion/erosion processes 38
General corrosion 39
Galvanic corrosion 39
Localised corrosion 40
Impingement and cavitation corrosion 40
Corrosion caused by polluted sea water 40
Bacterial corrosion 41
Sulphate reducers 41
Corrosion prevention 42
References 48

Chapter 4   Multi-stage flash distillation 50
Introduction 50
The submerged coil evaporator 52
Multi-stage flash principles 53
Stage number effect 56
Losses 61
Equilibration 61
Flash plant layout and construction 65
Heat recovery section 66
Heat rejection section 68
Pumps and auxiliaries 69
Air ejector 70
Brine heater 70
Plant operation 71
Summary 73
References 74

Chapter 5   Multiple-effect distillation 76
Introduction 76
VTE multiple-effect distillation—flow-sheet 77
Analysis of multiple-effect distillation 79
Fluted tubes 82
Flow distribution 86
Horizontal tube evaporator: multiple-effect evaporator (HTE) 87
Multiple-effect plant layout and operation 89
Scaling in VTE plants 91
Present position 91
References 93

Chapter 6   Miscellaneous processes   94
Introduction   94
Vapour compression distillation   94
Solar distillation   97
Principle of solar distillation   97
Still construction   98
Performance   103
Low temperature difference distillation   104
Controlled flash evaporation (CFE)   106
Non-metallic heat transfer surfaces   107
Summary   111
References   111

Chapter 7   Part A. Conjunctive use of desalination and conventional water resources   113
Preface by author   113
Introduction   113
The economics of conjunctive use   114
Conjunctive desalination   115
Conjunctive use of desalination with a short critical period reservoir   115
Conjunctive use of desalination with a long critical period reservoir   117
Conjunctive use of desalination under conditions of restricted plant operation   120
Conjunctive use with vapour compression distillation   121
The future potential of conjunctive desalination   122
The impact of low load factor operation upon desalination plant design   123
Corrosion   123
Product water quality   124
Plant life-time   124
Computational procedure   125
Generation of operating rules   126
Desalination plant size   128

Part B. The treatment of desalinated water for domestic use   129
Introduction   129
Corrosion mechanisms   129
Corrosion control   130
References   135

Chapter 8   Combined power and water production   136
Introduction   136
Combination plants   136
Nuclear power and desalting plants   138
Cost allocation   140
Gas turbine dual-purpose plants   143
Total Energy and postscript   145
References   147
Index   149

Notation

A	area of heat transfer surface m^2 (ft^2).
As	specific heat transfer surface area m^2 per kg/hr (ft^2 per lb/hr)
C	specific heat of liquid kJ/kg°C (Btu/lb °F)
h	heat transfer coefficient W/m^2 °C (Btu/ft^2/h °F)
L	latent heat of evaporation of water assumed throughout as 2.33 x 10^6 J/kg (1 000 Btu/lb)
n	number of stages/effects
n_r	number of heat recovery stages
n_j	number of heat rejection stages
P	pressure, N/m^2·Pa. (lb/in^2)
p_{sat}	vapour saturation pressure
Q	heat flux W (Btu/h)
r	brine circulation ratio
R	performance ratio
t	generalised fluid temperature °C (°F)
t_{bi}	brine stream temperature at inlet to stage i
t_{in}	brine inlet temperature to brine heater
t_{inlet}	generalised fluid inlet temperature for heat exchange calculations
t_{out}	generalised fluid temperature for heat exchange calculations
t_{max}	maximum temperature to which sea water is heated
ΔT	temperature difference, °C (°F)
T	generalised vapour temperature
T_{min}	blowdown temperature
T_s	vapour saturation temperature
$\Delta T \log$	logarithmic mean temperature difference
Δt_m	minimum stage temperature difference between vapour and recycle brine stream
Δ_{BPE}	boiling point elevation
U	overall heat transfer coefficient W/m^2°C, (Btu/ft^2 h °F)
p.p.m.	concentration parts per million
[ ]	molar concentration symbols

Additional nomenclature – Multi-stage flash evaporation

h brine depth m (or ft)
L_o initiation length m (ft)
n_e evaporation index
M_r brine recirculation mass flow rate – kg/h (lb/h)
M_d product mass flow rate
M_f feed mass flow rate
M_b brine blowdown mass flow rate
U mean bulk brine velocity m/sec (ft/sec)
U_c critical brine velocity m/sec (ft/sec)
X salt concentration
β fraction of equilibration
θ time

Additional nomenclature – Multiple-effect distillation

M mass flow rate, kg/h (lb/h)
fo feed stream prior to entry to effect 1
f feed stream to individual effect
d distillate produced in a specific stage
D total distillate leaving a stage
D' steam produced by distillate flashing
s steam stream
So feed steam to evaporator
Soo feed steam to preheater
k flashing constant $\Delta T_x/L$
ΔT_x intereffect temperature difference
ΔT_T evaporation temperature range
n number of effects
γ_T terminal temperature difference
α boiling point elevation

Note: Temperatures throughout the text will be referred in °C or °F. Absolute temperatures (K or R) will not be used as it is felt this confuses rather than enlightens.

Conversion table

Given a quantity in these units	Multiply by	To get quantity in these units
Pounds	453.59	Grammes
Kilogrammes	2.204 6	Pounds
Inches	2.540 0	Centimetres
Metres	39.370	Inches
Gallons (US)	3.785 3	Litres
Gallons (US)	231.00	Cubic inches
Gallons (US)	0.133 68	Cubic feet
Cubic feet	28.316	Litres
Gallons (Imp.)	1.2	Gallons (US)

1 m.g.d. = 4 546 m^3/day
£1 sterling = \$2.4 US

Notes:
1. All pence are new pence (p), i.e. £1 sterling = 100 p = \$2.4 US
2. All gallons are Imperial gallons.
3. All logarithms are to the base 'e' unless stated otherwise.

Preface

This book is intended to be an introduction to the principles of saline water distillation processes for water supply engineers, national resource planners, civil and mechanical engineers whose professional interests involve them in the consideration of potable water suplies from saline sources. It should also serve as a comprehensive introduction to the field for those who wish or need to know more about this growing area. It is not generally recognised that over 85 per cent of the world's installed desalting capacity utilises distillation. As yet no technically oriented book has appeared which devotes itself wholly to this process, indeed the less common methods of desalination such as reverse osmosis or freezing often have a preponderance of published material devoted to them at the expense of distillation. It is the author's aim to remedy this imbalance and provide a technically oriented readership with an introduction to the principles, design and operation of distillation plants. The energy resource implications of desalination are also considered.

Bibliographical references, listed at the end of chapters and referenced in text by numbers in [square brackets] are given only if the author feels they add to the understanding of the subject, since a review of the research literature is not one of the objectives.

Acknowledgements

The following persons, institutions and companies have rendered invaluable assistance in the preparation of this text. My thanks are due to all of them.

Aiton & Co. Ltd.
M. J. Burley
S. Chambers
Alan Dickson
Elsevier Publishing Company
A. Frankel
R. I. Hawes
Inst. of Mechanical Engineers
P. Mawer
Phys. Soc.
Porthan Limited
Royal Society of Edinburgh
H. C. Simpson
United Filters and Engineering Ltd.
U.K. Atomic Energy Authority
Water Research Association
Weir Westgarth Ltd.
F. C. Wood
Yorkshire Imperial Metals Ltd.

We are grateful to the following for permission to reproduce illustrations:

Fig. 1.1 M. King Hubbert, *Resources and Man*, Publication 1703, Committee on Resources and Man, National Academy of Sciences, W. H. Freeman & Co., San Francisco, 1969.

Fig. 1.2 *Report of the Royal Commission on Environmental Pollution*, Her Majesty's Stationery Office, 1971.

Fig. 1.3 *Reviews of Research on Arid Zone Hydrology*, UNESCO, 1953.

Figs. 1.4, 1.5, 3.1, 4.12, 5.1, 5.10, 6.2 United Kingdom Atomic Energy Authority.

Fig. 2.1 After M. Tribus and R. Evans, 'The thermo-economics of sea-water conversion', Report No. 62.53, University of California, February, 1963.

Fig. 3.6 H. C. Simpson and M. Hutchinson, Paper 43, 2nd European Symposium on Fresh water from the Sea, Athens, May 1967.

Fig. 3.7 Courtesy of Imperial Metals Ltd., Leeds.

Fig. 4.8 A. Porteous and R. Muncaster, *An analysis of equilibration rates in flashing flow with particular reference to the multi-stage flash distillation process,* Proc. 3rd Int. Symposium on Fresh Water from the Sea, Vol. 1, pp. 145–54, Djbrovnik, 1970.

Fig. 4.9 J. Winston Porter, 'Engineering for Pure Water, Part 1, Distillation', *Mechanical Engineering*, 90, No. 1 (January 1968), p. 19.

Figs. 5.8, 5.11, 5.13, 5.14 Aiton & Co. Ltd., Derby.

Fig. 5.12 C. Rhodes and K. B. Mills, 'The Gibraltar multiple-effect VTE falling film plant', Paper c/19/73, Conf. on Water Distillation, Institute of Mechanical Engineers, London, 1973.

Fig. 6.3 Courtesy of Buckley & Taylor Ltd., Oldham.

Figs. 6.9, 6.10 E. D. Howe *et al.*, 'Distillation schemes using low grade heat energy', 3rd Int. Symposium on Fresh water from the Sea, Djbrovnik, 1970.

Fig. 6.11 A. Kogan and A. Lavie, 'Conceptual design of a 50 MGD, MSF, direct contact condensation desalination plant', Communication from A. Kogan, *The Technion*, Israel, August 1973.

Figs. 7.1, 7.2, 7.3, 7.4, 7.5, 7.6 Dr. P. A. Mawer and Dr. M. J. Burley, Water Research Association.

Figs. 7.7, 7.8, 7.9 Courtesy of United Filters & Engineering Ltd., Mitcham.

Figs. 8.1, 8.2, 8.3, 8.4, 8.5 Wilfred H. Comtois, 'Combination of Nuclear Power and Desalting Plants', *Mechanical Engineering*, 89 No. 8. (August 1967), pp. 17, 19.

Figs. 8.6, 8.7 S. Chambers and F. C. Wood, 'Economic effect of water power ratios and the role of dual-process dual-purpose plants', *Journal of the British Nuclear Energy Society*, Vol. 7, No. 1, January 1968.

Fig. 8.8 N. O. Lier, 'Gas turbines prove effective for desalination', reprinted with permission from *Power*, November 1968, © McGraw Hill Inc., 1968.

Fig. 8.9 T. C. Carnavos, Small/medium dual-purpose plants, Proc. Conf. on Water for Peace, pp. 198–204, Washington, 1967.

Fig. 8.10 H. Kronberger and R. S. Silver, 'The role of desalination in water supplies', Paper P/74, Conf. on Water for Peace, Washington, 1967.

Chapter 1

A very general introduction

Man is by nature an exploiter of resources. There is nothing new in this remark. What we have to grasp is the scale of the exploitation and whether this will have a bearing on future technology. Desalination plants, in common with any other industrial installation, require raw materials for their construction and energy for their operation. The task of the engineer in today's society is not only the design of useful products or earning adequate returns on capital invested but to husband resources and deploy them so that future generations are well served also. The environment must be safeguarded for the self-same reasons. Because of this, distillation plants are not only considered at the design and performance level but also where pertinent the relevant environmental grounds will also be considered. For these reasons a very general look at some of the characteristics of life is called for including the general ramifications of the laws of thermodynamics – succeeding chapters will deal with their specific applications. The energy resource dilemma which we may face in the next generation is also considered.

Characteristics of life

Consider the following statements:

(a) *It is impossible to add to the material resources of the world.*
     (Resources fall into two categories, renewable and non-renewable – timber and water are examples of renewable resources; they are, of course, finite in their availability. Fossil fuels, copper, iron ore are examples of non-renewable resources. The time scale for a timber crop is roughly 40 years, oil reserves were formed over 600 million years.)

(b) *It is impracticable to dispose of waste materials outside the world and its envelope of air.*

Because of the implications of these statements nature has evolved a complex series of interrelationships whereby the essential materials required for life support are re-cycled and returned in a re-usable form. As an example, carbon is a principal element in the formation of natural compounds and as it is in finite supply a natural renewal cycle called the carbon cycle has evolved so that it is made available for re-use.

It is this re-cycling of the essential elements for life which is the keystone of our existence and is accomplished by the expenditure of energy. This is a basic law of life embodied in the laws of thermodynamics.

The first and second laws of thermodynamics

The laws of thermodymamics apply to all known phenomena.

The first law of thermodynamics

This law is a statement of the conservation of energy, i.e. energy can neither be created nor destroyed. Energy has many forms, e.g. electrical, chemical (the combustion of coal is a release of chemical energy), thermal nuclear, etc. The first law tells us that it is the total of energy in all its various forms which is a constant. Physical processes only change the distribution of energy, never the sum. However, the first law only tells us energy is conserved, it does not tell us the direction in which processes proceed. This is the province of the second law of thermo-dynamics.

The second law of thermodynamics

This law specifies the direction in which physical processes proceed. Its usual statements go along the lines that heat transfer takes place from a hot body to a cooler body or 'It is impossible to construct a system which will operate in a cycle, extract energy from a reservoir and do an equivalent amount of work on the sur-roundings'. Note that the opposite process, heat flow from a cold body to a hot body (the hot body becoming warmer the cold body becoming cooler) does not violate the first law. It does not occur because it violates the second law.

Thus the second law tells us, for instance, that concentrations will disappear, e.g. a sugar lump will dissolve in water, or in general that order becomes disorder. The general statement of the second law is that everything proceeds to a state of maximum disorder, and by implication is less useful in the event of the disorder occurring, e.g. a sugar lump is more useful when it isn't dissolved in the water. A kettle of boiling water is more useful than the same amount of water mixed in a bath of cool water. All these examples are manifestations of the second law. It applies to all biological and technological processes and cannot be circumvented.

The relevance of the second law for us is that the progression from order to a state of disorder demands the inevitable degradation of energy from the usable to the unusable. The first law says the total amount of energy in the universe is constant and the second law states that the fraction of energy available for use is constantly diminishing, i.e. 'the universe is running down' and man is contributing to this process. We consume ordered materials such as petroleum, coal, sugar, etc., to provide energy for our way of life. The end result is the degradation from high temperature to low temperature, and the second law tells us that more and more energy is becoming less and less usable. The second law requires that the fraction of energy available for use is continually being diminished.

This degradation of energy ultimately manifests itself in the production of thermal energy (heat) at relatively low and thus useless temperatures, e.g. the heat emission from a car engine exhaust, the hot water discharge from a power station, the heat of tyre friction on the road, the heat loss from the body, the decay of a rotting carcass are all forms of relatively unusable and degraded energy.

The laws of thermodynamics are often used to determine the ideal efficiency for a given process. All that need be noted is that for thermodynamic reasons alone, losses always occur. In the real world friction or its equivalent is always present so the losses are even larger than the theoretical ones.

Thus the laws of thermodynamics govern the behaviour of all living systems. They tell us why we need a steady input of energy to maintain ourselves, why, in order to have a cow gain 1 kg, it must eat much more than 1 kg in food. Why in order to distill 1 kg water from sea water the energy input required is much greater than the minimum energy of separation. Why both energy economy and low heat transfer surface costs are difficult to achieve, etc.

The laws of thermodynamics impose limitations on how far we can go, how far the earth will support species, and if the earth can only grow a certain amount of food (energy) then the laws impose restrictions on the number of mouths which can be fed.

Energy resources

The laws of thermodynamics also make it clear that all energy used on the face of the earth will eventually be degraded to heat at low temperature (by low is meant near the temperature of the surroundings). Thus just as a power station consumes fuel (an ordered structure) to produce thermal energy to make electricity, eventually the net result of the fuel consumption is to produce the equivalent amount of heat at low temperature. This implies that finite energy resources must not be squandered and the time scale of their exploitation has implications for designers if future scarcity means more expensive energy.

The scale of man's inroads on the earth's finite energy reserves is shown by Fig. 1.1 – estimates of the world crude oil production cycle using two values of Q_∞ – ultimate total oil production. Oil is a capital resource and is non-renewable except on a geological time scale, thus once these reserves are degraded to heat they are unrecoverable.

It is to be noted that two estimates of Q_∞ are used, the optimistic estimate of $2\,100 \times 10^{18}$ barrels shows that reserves will be virtually depleted by the year 2032 and on a less optimistic estimate near depletion will occur around 2010. These sobering projections can only be reflected by much higher prices for energy shortly as demand outstrips supply. For distillation purposes it emphasises the fact that energy cost estimates may not hold true for many plants and that during the projected lifetime the product water cost may rise considerably due to energy scarcity. The energy cost uncertainty will not apply to solar devices as they do not make use of a non-renewable energy supply but they as yet provide a very small

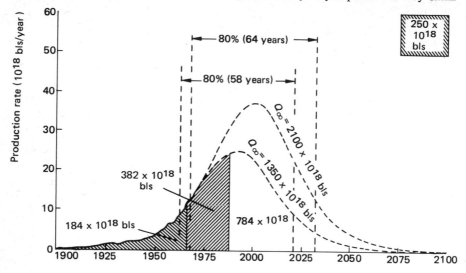

Fig. 1.1. Complete cycles of world crude-oil production for two values of Q_∞.

fraction of the total of desalted water. This text will concentrate on fundamentals and design principles and not the controversial area of costs as these are liable to many interpretations. The factors influencing costs will, of course, be discussed.

The hydrological cycle

Two of the principal factors governing development and growth are the availability of energy and potable water. However, the availability of water is governed by the hydrological cycle.

As with every other natural cycle there is a roughly constant amount in circulation at any one time, i.e. water is a finite resource (over 98 per cent at any one time is in the seas and much less than 1 per cent is actually available for human use). Some idea of the margins on which life support depends should be gleaned from this – the cyclic turnover cannot be increased as the solar energy input to it is constant, we can only utilise what we have more efficiently or augment supplies by desalination.

The first stage of the cycle is evaporation of water from the ocean. The evaporated water is then transported over land masses by air currents. Under favourable conditions precipitation will take place and rainfall in usable quantities deposited. Thus development can only take place where precipitation readily exceeds evaporation and usable water supplies become available by catchment or abstraction. The remainder of the precipitation then returns to the oceans by streams, thus completing the cycle. All that we do is temporarily catch some of the water at a suitable part in the cycle, use it for our purposes and return it as waste water to the cycle. The amount available for abstraction can never exceed the net precipitation and in fact never really approaches it. In the United Kingdom the volume of water abstracted by catchment schemes is approximately one-sixth of the total rainfall and to significantly increase this fraction will entail much greater cost than that at present incurred.

The United Kingdom water supply derives from the following sources: one-third comes from high ground, one-third comes from boreholes in aquifers (e.g. the London chalk basin supplies over $3\,000\,000$ m^3/day) and the other one-third comes from abstraction from rivers. Aquifer abstraction is now at its limits, i.e. it is not possible to sustain higher pumping rates. Also, new catchment areas on high ground are becoming scarcer, costlier and environmentally objectionable. The limits of conventional water supplies are being reached and it is from rivers that much of Britain's new water supplies will come – rivers which are the recipients of liquid wastes, sewage and other effluents. Sewage effluent is now a significant proportion of the volume of many rivers whose water is now used for public water supply as the demand presses on limited resources. We will return to this aspect later in the context of water re-use and desalination process selection.

Water supply and demand

As the demand for water increases so does the cost. In the non-arid zones increasing the supply can take the shape of more catchment schemes, aquifer abstraction, river abstraction and, of course, re-use. Figure 1.2 shows that the projected UK water demands will double in 25 years. (Some sources put the doubling time at 15 years.) Yet the UK has a relatively stable population whose doubling time is roughly 110 years. This tells us that per capita consumption is growing. The demands of industry and affluence in the UK mean that water supply is a critical factor for the growth of many areas. The experimental desalination of sea water by freezing has been considered for Ipswich. Reverse osmosis is being evaluated in various pilot projects. Thus even in relatively water-rich Britain desalination processes are under active consideration.

It is the world's arid and semi-arid zones that a real water need is evident. Some idea of the earth's arid and semi-arid areas is given in the UNESCO map, Fig. 1.3. It is in these places where there is little or no net precipitation that the need is greatest. What the arid zones quite often have is access to the sea or brackish water and thus some form of desalination can be practised.

Water quality standards

Water for domestic use should be free from pathogenic bacteria and disease and also comply with purity standards similar to those in Table 1.1.

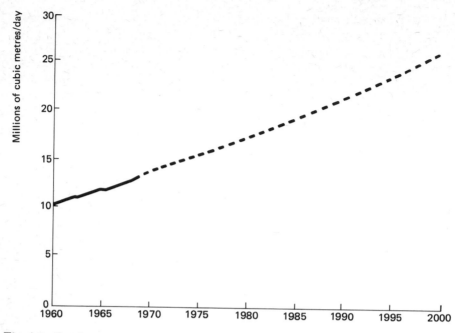

Fig. 1.2. Total water consumption 1960–69 with estimated requirements to year 2000 in England and Wales.

Table 1.1. World Health Organisation water purity limits

Substance	Maximum concentration permissible in public water supplies – p.p.m.
Carbon dioxide	20
Carbonates of sodium and potassium	150
Chlorides	250
Chlorine (free)	1.0
Copper	3.0
Detergents	1.0
Fluorine (as fluorides)	1.5
Iron	0.3
Lead	0.1
Magnesium	125
Nitrates	10
Phenols	0.001
Sulphates	250
Zinc	15
Total solids in suspension	500
Maximum NaCl	250

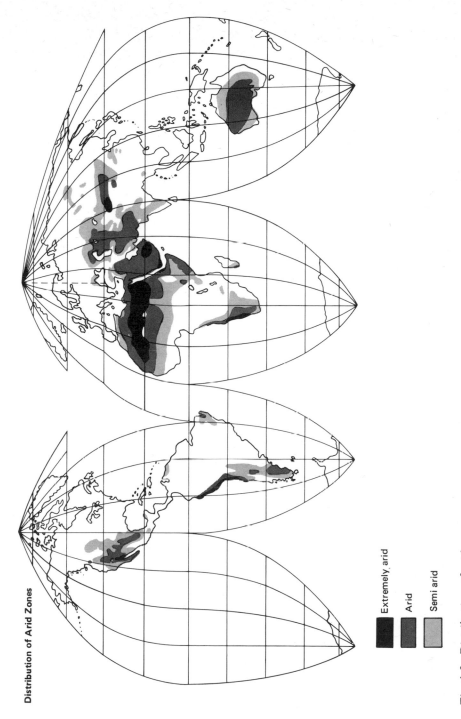

Fig. 1.3. Distribution of arid zones

The standards of Table 1.1 are often exceeded in arid zones were water supplies may contain in excess of 1 000 parts per million (p.p.m.) total dissolved solids. For most supply purposes 500 p.p.m. is taken as the maximum permissible upper limit. It is useful to put the upper limit of 500 p.p.m. in perspective. A typical stream in the UK can have 200–300 p.p.m. total dissolved solids prior to abstraction. After passing through the human organism and then to sewage treatment plants the sewage plant effluent can have another 200 p.p.m. dissolved solids added due to salt in the human diet, the breakdown of organic matter in plants, etc. Thus for a hypothetical case where a river has an initial total dissolved solid content of 200 p.p.m. and an abstraction rate of 50 per cent of the total flow is practised, on the return of the effluent stream to the parent, the total dissolved solids content will rise to 300 p.p.m. for once-through use. If the river serves several connurbations downstream each practising abstraction (at say 50 per cent of the flow rate) and adding an effluent with an additional 200 p.p.m. then only three abstraction cycles are possible before the 500 p.p.m. dissolved solid limit is met. This example shows the inevitable degradation which takes place during use. Demineralising processes may be necessary before repeated/or any abstraction can be practised on some UK streams. The River Trent is a good example – it receives sewage effluent from Birmingham plus industrial discharges. The net result is that the Trent at Nottingham is unfit to drink and a resource which could yield up to 1×10^6 m^3/day is currently unavailable – demineralising processes are being tested for suitability on the Trent water. Thus even in Britain water can be a resource which may require desalination techniques similar to the arid and semi-arid zones. The Colorado river is another example of the degradation which occurs in use, the US government has agreed to desalinate 4.546×10^6 m^3/day (10^8 gal/day) before it passes to Mexico to permit further usage.

Desalination processes and their applicability

The major desalination processes available are reverse osmosis, electrodialysis, freezing and distillation. The range of applicability of one process over the other is determined primarily by the salinity and composition of the feed-water.

Salinity is defined as the total amount of dissolved material in terms of kilogrammes of material per million kilogrammes of feed-water, i.e. parts per million (p.p.m.) of total dissolved solids. Chapter 3 discusses the composition of feed-water in detail, it suffices at the moment to say that a typical sea water has a salinity of 35 000 p.p.m. of which 30 000 p.p.m. is NaCl or salt. Now accepted potable water standards are a maximum of 250 p.p.m. as NaCl and a total dissolved solids content of 500 p.p.m. and these figures should preferably be lower.

Bearing this in mind if a process can only sustain a 90 per cent separation of total dissolved solids from a feed of 35 000 p.p.m. then the product water will have a dissolved solids content of 3 500 p.p.m. Water of this salinity is obviously not potable – this restriction applies to electrodialysis and currently to reverse osmosis but not to distillation as product purity can range from 1–100 p.p.m. from a feed of

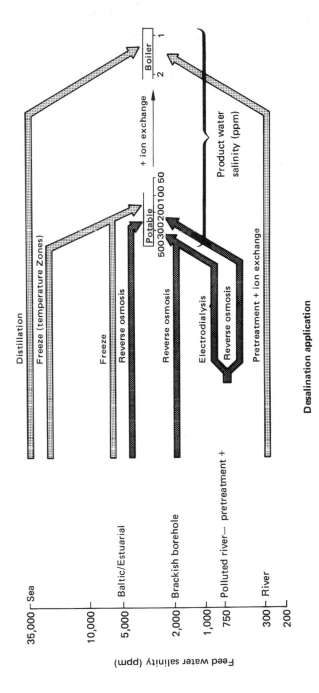

Fig. 1.4. Feed-water salinity and process applicability.

35 000 p.p.m., i.e. in order to be cost competitive with distillation (which implies a potable product) the feed-water for electrodialysis and reverse osmosis must be available with a total dissolved solids content much below that of sea water. Figure 1.4 illustrates this aspect in schematic form and the normal commercial feed-water

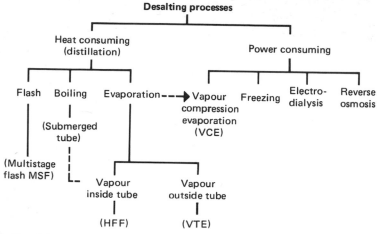

Fig. 1.5. Desalting processes.

ranges of all four processes. Figure 1.5 shows the basic desalting processes as being either heat consuming (distillation) or power consuming and categorises the variations in each class.

Process summary

Electrodialysis

Electrodialysis involves the passage of electric current through brackish or low salinity water in a chamber in which many closely spaced ion selective membranes are placed, thus dividing the chamber into compartments. The electric current causes the salts to be concentrated in alternate compartments with reduced salt content in the remainder. The constraints of electrodialysis are such that feed-waters up to 2 000 p.p.m. may be processed to give a product with 300 p.p.m. total dissolved solids. Where a high purity product is desired the costs may be unacceptable compared to water obtainable from distillation. A principal disadvantage of electrodialysis is that power consumption is proportional to total dissolved solids and its range of cost and performance effectiveness is with low salinity waters. Also high sodium chloride content in the feed-water will require a product purity of less than 250 p.p.m. which also adds to the cost – this restriction also applies to reverse osmosis.

Reverse osmosis

Reverse osmosis uses the reverse application of osmotic pressure. When salt water and fresh water are separated by a semi-permeable membrane, osmotic pressure causes the fresh water to flow through the membrane to dilute the saline water until osmotic equilibrium is established. Applying this in reverse, if a greater pressure is applied to the salt water then relatively pure water will pass through the membrane leaving a concentrated brine to be disposed of. The process technology is intimately bound up in the semi-permeable cellulose acetate membranes employed. Basically the rejection ratio, flux and membrane life are the principal factors. Usually high salt rejection is achieved at the expense of low flux and vice versa. The limits of 'once-through' operations are such that the approximate limit on the feed is 10 000 p.p.m. total dissolved solids content for high rejection membranes. The principal application may well be the treatment of industrial and sewage effluents or polluted rivers in the 750–10 000 p.p.m. range for either re-use or acceptable discharge criteria purposes. As with electrodialysis power consumption is proportional to total dissolved solids. The likely developments are a progression towards sea-water treatment but as yet no membrane has been developed which can sustain better than 99 per cent salt rejection which must be achieved before a potable water can be obtained from sea water with a total dissolved solids content of 35 000 p.p.m. The membranes are susceptible to bacterial fouling when applied to certain classes of effluent.

Freezing

The freezing of a salt solution causes crystals of pure water to nucleate and grow, leaving a brine concentrate behind. One commonly proposed freezing method is the use of a secondary refrigerant in which butane is evaporated in direct contact with sea water to remove the latent heat of crystallisation, as described below.

Sea water is precooled by heat exchange with the product water and waste brine streams. The sea water then enters the freezer where liquid butane is bubbled through the sea water. The butane vaporises and lowers the water temperature. This results in the formation of salt-free ice crystals in a more concentrated sea water or brine. Approximately one-half of the sea water is frozen into ice crystals. The ice-brine slurry is then pumped to a washer-melter. The slurry rises within the washer and the ice crystals are compacted into a porous bed of ice. The bed of ice is removed upward by a slight positive pressure caused by the brine flowing through the bed and outward through screens positioned near the middle of the column. The rising ice bed is washed with less than 5 per cent of the total product water. The ice is then removed by means of a mechanical scraper into the outer annulus; that is, the melter. The butane vapour, which contains the heat removed to form the ice, is compressed in the primary compressor and then introduced into the melter where it condenses on the ice. Heat is given up and the ice is melted. The condensed butane and the product water flow together to a decanting unit where the two liquids are separated.

From the decanter, the product water leaves the process and the liquid butane is recycled back to the freezer. Butane vapour not required for ice melting is further compressed by the secondary compressor and then condensed in the butane condenser, which is cooled by sea water. The liquid butane is re-cycled to the freezer.

Distillation

Distillation involves the boiling or evaporation of sea water to form water vapour which is condensed to yield a salt-free stream. Energy requirements are virtually independent of the feed-water salinity and product purity of less than 50 p.p.m. can readily be achieved. The principal choices of plant are Multiple Effect evaporation (ME) and Multi-Stage Flash distillation (MSF). The inclusion of 'multi' means that a cascading effect is employed whereby multiple re-use of the energy content of the heating steam is obtained, i.e. a succession of evaporation condensation processes is performed within the one plant. The main multiple effect evaporation process is the Long Tube Vertical process (LTV), where the steam generated in one effect condenses on the outside of long vertical tubes in the next effect thus evaporating more water from a film of saline water which runs down the inside of the tube. This process has strong growth potential as the development of extended heat transfer surfaces is likely to give lower water costs than those of comparable MSF plants. Chapters 4 and 5 discuss the MSF and ME processes respectively.

In MSF distillation, sea water is heated under pressure to an elevated temperature in the range 90°–125°C but no boiling is allowed to take place. The hot sea water or brine is then discharged through a series of stages each at a reduced pressure compared with the previous stage. Evaporation or flashing of vapour takes place from the hot brine stream in each stage in an attempt to obtain equilibrium with the vapour conditions prevailing in each individual stage. In this way, a portion of the flashing stream is evaporated in each stage. The controlling parameters on costs and output are: the temperature drop allowed in each stage; the overall flash range, i.e. the difference between brine inlet temperature to the first stage and discharge temperature at the last; and lastly, the stage heat transfer coefficients. All these factors and others will be discussed in detail later. Let it suffice to say that over 85 per cent of the world's installed desalting capacity is provided by distillation plants with MSF currently in total dominance. Sight should never be lost of this dominance, distillation is a proven process on almost all feed waters – electrodialysis, while a proven process, will probably never operate on sea water and reverse osmosis is in its technological infancy with respect to tackling sea water also.

Other processes have been mooted, e.g. freezing, hydrate formation, ion exchange, but unless and until these are proven in the market place in free competition with distillation plants of the same capacity, little attention can be paid to them by those who have the task of providing reliable water supplies now and in the foreseeable future.

This book will now deal with principles and practices of distillation and the MSF and ME processes. Consideration will also be given to the as yet comparatively minor processes of solar and vapour compression distillation as these fill a niche in the distillation spectrum.

The integration of desalination plants with conventional water supplies will also be discussed and the role of single- and dual-purpose plants reviewed. Above all the aim is to give information in a straightforward manner for those who have the difficult task of choosing between competing designs or initiating specifications for desalination plants. References to the literature are provided not for the sake of erudition but because the author feels they impart useful information. The underlying principles of the laws of thermodynamics will be evident throughout.

The author has deliberately avoided the tendentious field of water costs. Where cost information is desired reference can readily be made to the relevant Office of Saline Water Publications for an introduction to standardised cost formats and individual manufacturers or consultants for order of magnitude costs.

In order to understand the constraints governing the design of distillation plants, Chapter 2 gives some thermodynamic and heat transfer fundamentals. Chapter 3 deals with the chemistry of sea water and the prevention of scale and corrosion which underly the design and operation of most distillation processes.

Chapter 2

Fundamentals

Introduction

An understanding of distillation plant design requires a knowledge of certain fundamental aspects such as vapour–liquid equilibrium, heat transfer coefficients, etc. This chapter lays no claim to completeness of the subject material; it is included so that a common basis for the discussion of the process technology is available.

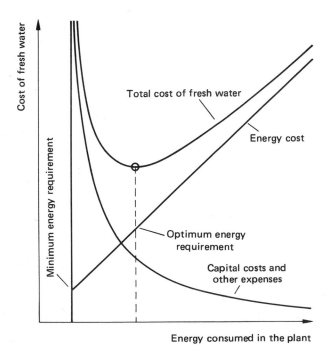

Fig. 2.1. Qualitative relationship between the minimum and the optimum energy requirements of a sea-water conversion plant.

The principal factors governing distillation plant design are the laws of thermodynamics and their application in practice. It is all very well to state that there is a minimum work of separation of fresh water from sea water and infer that major departures from this theoretical attainment occur, in practice this tells us very little. It is a law of nature that irreversible losses occur and distillation plants are no exception. A better insight is gained by looking at these irreversible processes and seeing what the designer can do in practice to minimise them consistent with obtaining a minimum water cost.

Figure 2.1 shows the qualitative relationship between the minimum and the optimal energy requirements of a distillation plant after Tribus and Evans. [1] The total cost of the distillate is given by the sum of the costs for energy plus capital plus any other costs. Figure 2.1 is very illuminating, it shows that to obtain the minimum energy requirement infinite capital cost is required which in turn dictates infinite water costs. At the other extreme low capital costs imply high energy consumption with again a commensurable high cost of fresh water. Somewhere between the two extremes, there is an optimum struck between the *level of economics and technology* (I am indebted to Myron Tribus for this caveat on an engineer's version of optimum).

We can now look at some of the factors which introduce irreversibility and thus influence costs of distillation plants. The first and largest costs component in any distillation plant is the heat transfer surface area.

Heat transfer

Heat transfer occurs because of a temperature difference. In distillation plants there is the requirement to evaporate from a liquid phase and condense the vapour to form the distillate. The condensation is often achieved by enclosing the cool sea-water feed to the evaporation section in a tube envelope and recovering the latent heat of vaporisation, thereby partially heating the feed-water.

Thus heat recovery is obtained, the heat transfer taking place from the vapour space through the tube wall and attendant condensate film and other impedances to heat transfer ultimately to the brine in the tubes. However, merely stating this conveys no impression of the magnitudes of the resistances to the heat transfer process.

Figure 2.2 shows, in conjunction with Fig. 2.3, the mechanism of heat transfer from the vapour space in an evaporator to the brine in the tubes. The tube configuration is horizontal and the various resistances to the transfer process are characterised by heat transfer coefficients with the respective values as follows: (For a discussion on heat transfer principles and units employed see Kern [2] or bird, Stewart and Lightfoot [3].)

Interfacial heat transfer coefficient
$$h_i = 113\ 400\ \text{W/m}^2\ °\text{C}$$
$$(20\ 000\ \text{Btu/ft}^2\ \text{h}\ °\text{F})$$

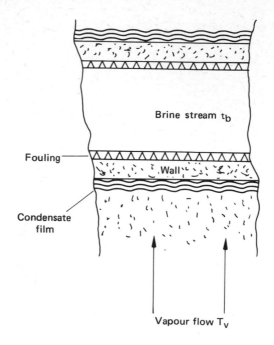

Fig. 2.2. Heat transfer through a horizontal condenser tube wall.

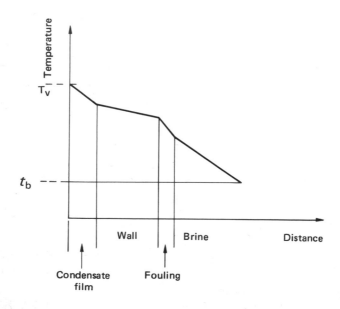

Fig. 2.3. Qualitative temperature profile for Fig. 2.2.

Condensate film heat transfer coefficient
$$h_1 = 11\,340\text{--}14\,200 \text{ W/m}^2 \text{ }^\circ\text{C}$$
$$(2\,000\text{--}2\,500 \text{ Btu/ft}^2 \text{ h }^\circ\text{F})$$

Tube wall heat transfer coefficient
$$h_t = 68\,000 \text{ W/m}^2 \text{ }^\circ\text{C}$$
$$(12\,000 \text{ Btu/ft}^2 \text{ h }^\circ\text{F})$$

Brine heat transfer coefficient
$$h_w = 13\,000 \text{ W/m}^2 \text{ }^\circ\text{C}$$
$$(2\,300 \text{ Btu/ft}^2 \text{ h }^\circ\text{F}).$$

The effect of fouling is allowed for by the use of a fouling factor. h_i is a property of the liquid-vapour system and the magnitudes of h_1, h_t, h_w, cannot be increased without considerable expense or major changes in technology. Briefly:

(a) The interfacial coefficient h_i results from molecular interchange considerations and is therefore fixed.
(b) The liquid film coefficient results from the condensate drainage and is fixed by the mechanics of the condensation process.
(c) The tube heat transfer coefficient is determined by the thermal conductivity of the tube material and the tube wall thickness.
(d) The brine heat transfer coefficient is a function of the fluid properties and the velocity of flow in the tubes.

For most purposes the velocity of flow is fixed by pumping power requirements as the greater the brine velocity, the greater the frictional dissipation and therefore more power is required, i.e. an increase in the brine heat transfer coefficient is bought at the expense of pumping power (plus tube erosion at higher velocities for some heat transfer materials). It is seen that the main heat transfer components are relatively fixed. The overall heat transfer coefficient is obtained by the standard heat transfer relationship in equation (2.1) below (see for example Kern [2]).

$$\frac{1}{U} = \frac{1}{h_i} + \frac{1}{h_1} + \frac{1}{h_t} + \frac{1}{h_w} + \begin{matrix}\text{fouling}\\\text{factor}\end{matrix} \qquad (2.1)$$

where the fouling factor is introduced to allow for the effects of scale build-up in the tubes or sludge deposition and is discussed in detail shortly. Substituting the respective values in equation (2.1), and using the Btu/ft^2 h $^\circ$F values for illustration only:

$$\frac{1}{U} = \frac{1}{20\,000} + \frac{1}{2\,000} + \frac{1}{12\,000} + \frac{1}{2\,300}$$
$$= (5 + 50 + 8.33 + 43.5)\,10^{-5}$$
$$= 106.83 \times 10^{-5}$$

Thus $\qquad\qquad U = 940 \text{ Btu/ft}^2 \text{ h }^\circ\text{F } (5\,330 \text{ W/m}^2 \text{ }^\circ\text{C}).$

The fouling factor used can vary in practice from 0.000 5 for clean tubes to 0.001 for adverse scale control conditions. Thus the effective heat transfer coefficient can range from 3 660 to 2 750 W/m² °C (646 to 485 Btu/ft² h °F). The area required to transfer a given heat flux is given by

$$A = \frac{Q}{U\Delta T \log} \tag{2.2}$$

where Q is the required heat flux, ΔT log is the logarithmic mean temperature difference which in this case is a function of the vapour temperature, brine temperature at tube inlet and brine temperature at tube exit, and U is the overall heat transfer coefficient as discussed above.

Commonly in heat exchanger design, the tube side fluid temperature and mass flow rate is known. Thus an energy balance on the heat transfer system whereby heat flux Q is transferred to tube side fluid, gives the exit temperature, i.e.

$$Q = MC(t_{out} - t_{inlet}) \tag{2.3}$$

where M is the mass flow rate and C the fluid specific heat, and t_{out} and t_{inlet} are generalised fluid outlet and inlet temperatures respectively. Thus t_{out} may be found and hence ΔT log calculated. Now in an MSF plant the recirculation rate is fixed. Thus t_{out} at each stage is proscribed. Heat rejection is however accomplished by a separate circuit and as a greater cooling water flow rate is used, t_{out} is reduced for a given heat flux. Hence the heat transfer area for the rejection section may be changed by varying M as shown by equations (2.2) and (2.3).

Now if Q is constant, U and ΔT log can be design variables. Let us assume ΔT is fixed and U can vary over the range 2 750 to 3 660 W/m² °C, then the area A for U = 2 750 W/m² °C is 31.5 per cent greater than that for U = 3 660 W/m² °C. Thus the maintenance of low fouling factors is of importance in plant operations. Chapter 3 discusses this and other aspects of scale prevention. The overall heat transfer coefficient U is seen to have well prescribed limits. Thus the designer has to manipulate the mean temperature difference ΔT to his advantage. However, there are constraints on ΔT as well which are discussed in Chapters 4 and 5.

Heat transfer surface and performance ratio

The schematic arrangement in Fig. 2.4 shows the principle of flash distillation. Concentrated sea water or brine is heated to just below its boiling point at pressure P_{max} and temperature t_{max} in the brine heater. The brine then flows to the flash chamber which is at pressure P_1 which is less than P_{max}. The pressure reduction causes flashing or instantaneous evaporation to take place until the brine is in thermodynamic equilibrium with the vapour conditions at $P_1 T_1$ (where t denotes brine temperature, and T denotes vapour temperatures, t_{bi} is the re-cycle brine temperature at stage 1 inlet, t_{in} is the re-cycle brine temperature at stage 1 exit i.e. the brine heater inlet temperature).

The flashing vapour is condensed and gives up its latent heat to the inlet sea water in the condenser tubes. The latent heat of evaporation is thus recovered and re-used. As the flash chamber runs at steady state the evaporation rate is constant and heat flux Q_1 to the condenser tubes is constant, Now, as we have seen the stage heat transfer coefficient U is a constant, then if a low energy input into the brine heater is desired t_{in} must be as near as practicably possible to t_{max}, so that the

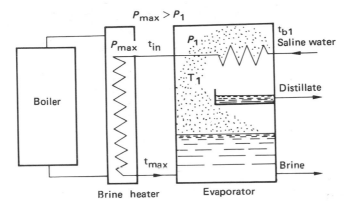

Fig. 2.4. Flash distillation (brine heater and first stage).

heating steam has to provide a small temperature difference $(t_{max} - t_{in})$ only. In order to do this, the brine inlet temperature to the feed heater must approach the vapour temperature T_1 in the first flash chamber. Inserting some typical values on these temperatures we have, using the vapour temperature T_1 as datum, that the inlet temperature difference between brine and vapour, i.e. $(T_1 - t_{bi})$ can be 4.16°C (7.5°F) and the terminal or minimum temperature difference between brine and vapour, i.e. $(T_1 - t_{in})$ is 2.78°C (5°F), i.e. the brine temperature in the condenser tubes has increased by 1.39°C (2.5°F) from the recovery of the latent heat of condensation of the vapour.

The logarithmic mean temperature difference for a condenser is calculated as follows (see Kern [2] for details).

$$\Delta T \log = \frac{t_{in} - t_{bi}}{\log\left(\dfrac{T_1 - t_{bi}}{T_1 - t_{in}}\right)} = \frac{1.38}{\log\left(\dfrac{4.16}{2.78}\right)} \tag{2.4}$$

$$= 3.4°C \ (6.1°F)$$

What this means is shown in the temperature plot of Fig. 2.5. The varying temperature difference may now be replaced by the logarithmic mean temperature difference of 3.4°C. These figures apply to an MSF plant with a performance ratio of 11 and a flash range of 55.5°C (100°F), i.e. the brine flashes over this total temperature range in 40 stages each with a 1.39°C (2.5°F) temperature drop. Performance ratio is defined as the number of pounds of distillate produced per pound of steam used in the brine heater, i.e. pounds distillate produced per 1 000 Btu steam consumed (kg product per 2.33×10^6 J steam consumed to keep comparison identical with the literature, where L is commonly taken as 1 000 Btu/lb.)

Comparable figures for a plant with a performance ratio of 5.5 are

$$(T_1 - t_b) = 7.5°C \ (13.5°F)$$
$$(T_1 - t_{in}) = 5.1°C \ (9.2°F)$$

with corresponding logarithmic mean temperature difference of 6.2°C (11.2°F). Thus the plant with the lower performance ratio will have a 45 per cent reduction in heat transfer area compared with the high performance ratio plant (given that all other factors are the same for both plants). This illustrates the relationship between energy costs and capital costs. A high performance ratio plant would be specified for a location where fuel is expensive whereas the low performance ratio plant would be used where fuel is very cheap, e.g. in Kuwait.

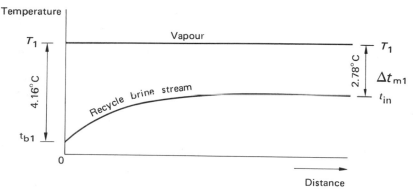

Fig. 2.5. Typical temperature profiles for brine and vapour streams in a heat recovery stage of an MSF plant (symbols used refer to first stage of the plant).

Many other factors interplay in distillation plant heat transfer considerations especially in those plants where a high performance ratio is required and low mean effective temperature differences are employed. The major source of temperature reduction or thermal energy degradation, as this is what a temperature reduction implies, are boiling point elevation and pressure drop losses respectively.

Boiling point elevation

The commonest was of explaining boiling point elevation is to imagine two insulated containers A and B as shown in Fig. 2.6. A contains sea water and B pure water. The sea water temperature is t_A, and The pure water temperature t_B, the vapour temperatures are equal and the vapour space between the two connected by a compressor. For equal temperatures in both A and B the vapour pressure in B is greater than that of A, i.e. $P_B > P_A$. Thus in order to prevent vapour transport from B to A the compressor must equalise the pressure. If the insulation is removed and temperature variation allowed t_A must be increased above t_B so that $P_A = P_B$. In other words the lower vapour pressure for the sea water manifests itself as a boiling point elevation which is a function of the dissolved solids concentration.

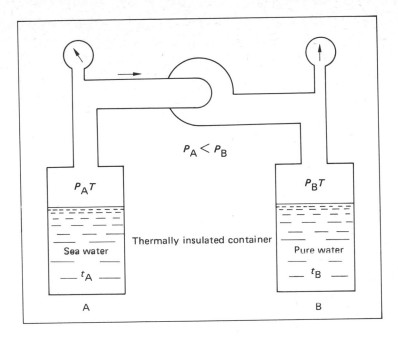

Fig. 2.6. Conceptual device for the measurement of boiling point elevation.

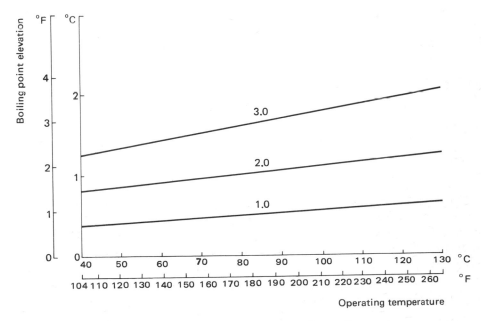

Fig. 2.7. Boiling point elevation for various concentrations of sea water.

Figure 2.7 shows typical boiling point elevations for various concentrations of sea water over the normal working range of sea water evaporators. For brine of 2 × sea water concentration at 88°C (190°F) and boiling point elevation is 1.11°C (2°F). What this means to the evaporator designer is that the water vapour evolved from brine at 88°C (190°F) in the flash chamber will have a temperature of 86.89°C (188°F) which is a serious loss for high performance ratio plants. Thus in the first stage of the evaporator of Fig. 2.4, the flashing brine exit temperature t_1 is 88°C and the vapour temperature $T_1 = 86.89$°C just above the brine surface.

Pressure-drop losses

However, other losses occur. A principal one is pressure-drop loss which takes place wherever the vapour encounters an obstruction in its path. The first and most severe obstruction in terms of pressure-drop loss occurs in the separators whose function is to disengage entrained droplets of brine which would otherwise affect product purity. Wire mesh demisters are commonly used for this process and they impose a pressure drop on the vapour stream. Corresponding to the pressure drop there is a saturation temperature drop when the vapour passes through the demister. A similar temperature drop is encountered when the vapour passes through the condenser tube banks. When the boiling point elevation and pressure drop effects are taken into consideration a total temperature loss c. 1.39°C (2.5°F) may be obtained. The correct assessment of these losses is of great importance in high-performance ratio evaporators.

Hydrostatic head effects

Temperature losses have so far been discussed in the context of MSF. However, in one form of ME distillation a major thermodynamic loss is introduced by hydrostatic head effects. Consider the single-effect submerged coil evaporator shown in Fig. 2.8. Heating steam is sent through the coil and causes evaporation to take place. However, the steam temperature required for evaporation must be such that not only are the temperature losses of both boiling point elevation and pressure drop overcome but also the effect of hydrostatic head. As the heating tubes are submerged in the brine the static head effect raises the temperature at which the brine will boil at the tube surface. This can be considerable, e.g. in the low-pressure end of ME evaporators a depth of submergence of only 2.5 cm can require an extra 0.83°C (1.5°F) before boiling begins. If a tube nest is 50 cm deep an additional 16.5°C is required if evaporation is to take place throughout the pool. Hydrostatic head effects introduce an additional penalty on the design of submerged tube evaporators, and are mainly responsible for the demise of this process (for an excellent discussion of the rise of MSF and the decline of submerged tube evaporator plants see Silver [4]).

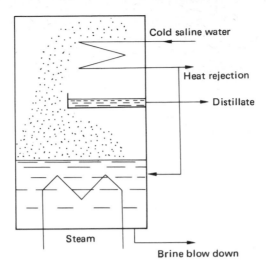

Fig. 2.8. Single-effect submerged coil evaporator.

The second law of thermodynamics

This has been discussed in Chapter 1 but a brief restatement of its implications in the desalination context is worthwhile. It is obvious that any desalting process consumes energy well in excess of the theoretical minimum. Some desalination processes use mechanical energy or electrical energy whereas distillation uses thermal energy. The maximum conversion of thermal to mechanical energy is given by the Carnot factor $(T_1 - T_2)/T_1$, in which T_1 and T_2 are the absolute temperatures of heat source and sink respectively. In practice this is typically 35 per cent for a modern power station. However, while mechanical and electrical energy can be completely transformed these forms of energy usually have their origins in a source of thermal energy, i.e. the use of mechanical or electrical energy is therefore inherently more expensive than thermal energy.

It is not often relevant to talk of the quantity of energy consumed per unit of water produced as a basis for process comparison. What must be compared is the energy cost per unit of water, e.g. reverse osmosis will consume electrical power whereas distillation uses thermal, but one kWh of electrical energy is much more expensive than one kWh of thermal energy (approximate ratio of 4 : 1). A distillation plant may therefore have a specific energy consumption of three times a reverse osmosis plant and still be competitive on energy cost terms.

When comparing plants of the same type, comparisons can of course be made in terms of energy consumption but sight must never be lost of the fact that it is product cost which matters and the lowest product cost for any given condition is a function of both energy and capital costs. Energy costs can be reduced in distillation plants at the expense of increased capital expenditure and a trade-off is made between the two – a theme which subsequent chapters will expand.

Flash range

The flash range of a MSF plant is the temperature range over which the plant operates. For any thermodynamic device which utilises a source at temperature T_{max} and a sink at temperature T_{min} the maximum possible temperature range over which the device can operate is $T_{max} - T_{min}$. For a distillation plant, T_{min} the sink temperature is fixed by location, e.g. the sea (and in practice from second-law considerations heat rejection is always at a temperature above T_{min}) thus the discharge temperature is fixed at $t_{discharge} > T_{min}$. The upper temperature for the flash plant is t_{max} as shown in Fig. 2.4 (p. 19). This has restrictions placed on it by scale prevention considerations and hence the flash range in an MSF plant is restricted by current technology to a maximum of 89°C (160°F). In order to make 1 kg of distillate by evaporation in a flash evaporator the flashing brine stream must circulate a much greater mass flow rate in the plant than the distillate output. A simple calculation will suffice. Assume that latent heat of vaporisation (L) of the flowing brine is 2.33 x 10^6 Joules/kg (1 000 Btu/lb) and a flash range of 55.5°C (100°F). Then the brine recirculated per kg of distillate is

$$r = \frac{L}{C(\text{flash range})} = \frac{L}{C(t_{max} - t_{discharge})}$$
$$= \frac{2.33 \times 10^6}{4\ 200\ (55.5)} = 10$$

where C the specific heat of brine is taken as 4 200 J/kg°C (1 Btu/lb °F).
   For a flash range of 89°C (160°F)

$$r = \frac{2.33 \times 10^6}{4\ 200(89)} = 6.25$$

thus the flash range is an important parameter in plant design which is also confined to a well-defined range by the second law of thermodynamics on the one hand and scaling limitations on the other.
   In the next chapter the mechanisms of scale and corrosion are examined as these are common to all forms of distillation and fix the permissible ranges of temperature and sea-water concentrations at which plants can operate.

References

1. Tribus, M. and Evans, R. 'The thermo-economics of sea water conversion', University of California, Report No. 62. 53, February 1963.
2. Kern, D. Q. *Process Heat Transfer*, McGraw-Hill.
3. Bird, R., Stewart, B. and Lightfoot, E. N. *Heat, Mass and Momentum Transfer*, Wiley 1960.
4. Silver, R. S., A review of distillation processes for fresh water production from the sea. Sanderwick and Dechema, Monograph Bond 47, 1962.

Chapter 3

The control of scale and corrosion

Introduction

The prevention of scale and corrosion, or more strictly the control of scale and corrosion (as quite often neither can be completely prevented economically) is of great importance in saline water distillation plant operation. The control of scale is discussed first.

It has been shown in Chapter 2 that the overall heat transfer coefficient for water flowing in a clean condenser tube at a velocity in the conventional range of 2-3 m/s is of order 5 678 W/m^2 °C (1 000 Btu/ft^2 h °F). When fouling occurs as typified by the inclusion of a fouling factor in equation (2.1), the overall coefficient can drop to 2 750–3 660 W/m^2 °C (485–646 Btu/ft^2 h °F) for fouling factors of 0.001 and 0.000 5 which respectively span the range for saline water distillation plants. This reduction in heat transfer coefficient is serious, using the heat transfer coefficients quoted above, if the fouling factor drops from 0.000 5 to 0.001. This means that the plant requires roughly 40 per cent more heat transfer surface for the same heat load. As a typical 4 546 m^3/day (1 m.g.d.) MSF plant may have 7 500 m^2 (80 000 ft^2) of heat transfer surface which at say £30/m^2 represents a capital cost of £225 000, then a 40 per cent increase in surface area, if design is based on a fouling factor of 0.001, is not to be treated lightly. The fouling factor is principally influenced by the build-up of scale or sludge on the heat transfer surfaces of the plant. Scale is defined as a hard crystalline deposit which adheres to the heat transfer surfaces and which requires physical methods such as chipping or drilling for its removal, whereas sludge is a soft silt type material which may remain in suspension or be deposited but does not firmly adhere to the heat transfer surfaces and its removal may be effected by increased water velocity or brushing. Scaling then, whose occurrence is a function of the properties of the raw feed, is an important factor in plant operation.

Composition and properties of typical raw feed-water

The composition and properties of the raw feed are extremely important in several respects. The principal ones are the treatment necessary to prevent scale and the determination of the economic boundaries which makes one purification process more economical to run than another. As far as distillation processes are concerned, the raw feed-water can vary within quite wide limits as the cost of the product is not nearly so dependent on the salinity of the feed or the product purity as in other processes.

The ideal raw feed-water would be one which produced no scale on the heat exchanger surfaces independent of temperature or concentration. In practice the actual raw feed available can vary widely in composition and scaling properties depending on its source which is usually from the sea or a bore-hole. Sea water is to be preferred due to its local constant composition in spite of high salinity. Bore-hole or brackish waters may have only one-third the salinity of sea water but may contain a very high concentration of calcium salts thus increasing the risk of severe scaling.

In order to quantify sea and brackish waters the following values in Table 3.1 will serve as rough guides.

Table 3.1.

Source	Salinity Total dissolved solids (p.p.m.)
Potable well water	300–500
Approximate limit for irrigation	1 000
Brackish well water	1 500–6 000
Baltic Sea	3 000–4 000
Arabian Gulf	44 000
'Typical' sea water	35 000

The total salinity is important not only from scale control considerations but also from a process selection point of view. At salinities below approximately 7 000 p.p.m. electrochemical processes such as electrodialysis, ion exchange or reverse osmosis may become viable alternatives to distillation processes. However, as power consumption or energy input in the electrochemical processes is roughly proportional to the salinity, a point is reached where distillation-type processes dominate where energy input is virtually independent of salinity. For plants above 2 265 m^3/day (500 000 gal/day) capacity, the lower energy requirements above a salinity of approximately 7 000 p.p.m. coupled with lower capital costs virtually ensure the installation of a distillation type plant.

In order to understand the mechanisms of scaling, a typical sea water will be considered in detail. The composition of the typical raw feed is given in Table 3.2 in terms of total dissolved solids g/kg and also molar concentration in g mol/l.

Grunberg [1] and Emmerson [2] give particulars of other sea-water compositions and detailed measurements of the various physical properties. For distillation processes the bicarbonate content of the feed is the most important as it is

Table 3.2. Composition of a typical sea water

	g/kg	g/l	g mol/l
Total salts	35.1	36.0	
Sodium (Na)	10.77	11.1	0.482
Magnesium (Mg)	1.30	1.33	0.055 4
Calcium (Ca)	0.409	0.42	0.010 5
Potassium (K)	0.388	0.39	0.010
Chloride (Cl)	19.37	19.8	0.558
Sulphate (SO_4)	2.71	2.76	0.028 8
Bromide (Br)	0.065	0.066	0.000 8
Bicarbonate (as HCO_3)	0.149	0.152	0.002 5

responsible for the majority of scaling, corrosion and non-condensable gases as discussed by Hillier. [3]

Two types of scale can form in distillation plants:

(a) Alkaline, i.e. those resulting from the decomposition of the bicarbonate content of sea water.

(b) Calcium sulphate scale.

We shall first discuss the formation and prevention of the former and then proceed to the latter.

Alkaline scales

The alkaline scales occur when the bicarbonate ion breaks down on heating, resulting in the following reaction

$$2H\,CO_3^- \rightleftharpoons CO_2^\uparrow + CO_3^{--} + H_2O \tag{3.1}$$

The carbonate ion concentration is a maximum when all the bicarbonate has been converted to carbonate by loss of CO_2.

The carbonate ion can now react in either of the following ways

$$Ca^{++} + CO_3^{--} \rightleftharpoons Ca\,CO_3^\downarrow \tag{3.2}$$

or

$$CO_3^{--} + H_2O \rightleftharpoons CO_2^\uparrow + 2OH^- \tag{3.3}$$

The hydroxyl ion can react with the magnesium ion present to form magnesium hydroxyl scale as in equation (3.4) below.

$$Mg^{++} + 2OH^- \rightleftharpoons Mg(OH)_2 \tag{3.4}$$

Thus the original bicarbonate content will, when heated, give an equivalent concentration of carbonate or hydroxyl ions – the respective concentrations which can exist in solution being governed by the solubility product of $CaCO_3$ and $Mg(OH)_2$ respectively. [3]

At temperatures below approximately 82°C (180°F) equation (3.2) predominates and $CaCO_3$ scale deposition is a likely outcome. Above 82°C (180°F) equation (3.3) predominates and hydroxyl ion formation is favoured leading to $Mg(OH)_2$ formation.

$CaCO_3$ and $Mg(OH)_2$ are both virtually at saturation concentration in ordinary sea water. As both constituents exhibit reverse solubility, i.e. the equilibrium

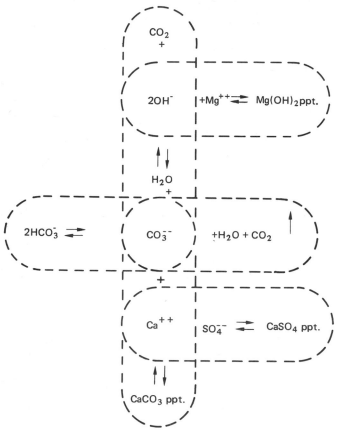

Fig. 3.1. Principal scale-forming reactions for sea water.

amount that can be retained decreases on heating once the saturation limit is reached. Figure 3.1 summarises the principal scale-forming reactions for sea water.

It is seen that for most distillation processes measures must be taken to suppress alkaline scale formation. Standiford and Sinek [4] and Langelier *et al.* [5] have studied the solubility of $CaCO_3$ and $Mg(OH)_2$ in sea-water brines of varying concentration and temperature.

Figure 3.2 shows the results of Standiford and Sinek for determining the solubility limits of $CaCO_3$ $Mg(OH)_2$. The results are presented as a plot of concentration factor against a base of pH and temperature. Thus if the pH and temperature of the sea-water brine stream are known a vertical line drawn to meet the appropriate concentration factor will give the coordinates for determining whether the solution under consideration is saturated or unsaturated with respect to $Mg(OH)_2$. $CaCO_3$ scale is also a function of the total alkalinity (milligrams $CaCO_3$ per litre). The correct saturation line for $CaCO_3$ is determined by the alkalinity of the solution. The no-loss line is the $CaCO_3$ saturation line for sea-water

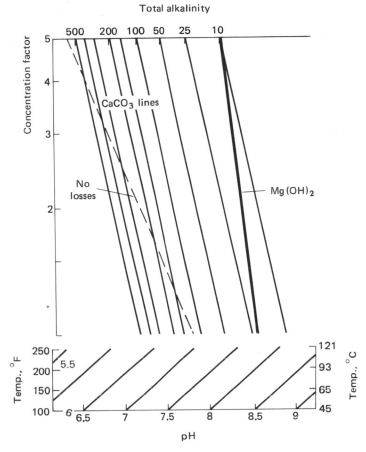

Fig. 3.2. Solubility of $CaCO_3$ and $Mg(OH)_2$ for varying sea-water concentrations and temperatures.

brines with no losses of alkalinity obtained by plotting total alkalinity variation with concentration factor for untreated sea water. In most plants the combination of both temperature and concentration factor at an untreated sea-water pH of eight will result in at least $CaCO_3$ scale formation and if a temperature of 82°C (180°F) is to be exceeded at any concentration factor and pH of 8 $Mg(OH)_2$ scale will also form. As pH is an independent variable, it may be reduced by hydrogen ion addition to the feed and enable the plant to operate free of alkaline scale. This is one method of scale control and is one of the following five main classes.

Methods of scale control

1. Mechanical or chemical means of loosening deposited scale.
2. Seeding to establish preferential growth on an easily removed surface.
3. Ion exchange.
4. Addition of proprietary scale inhibiting compounds.
5. pH control.

Although the most commonly-used methods for large distillation plants are methods 4 and 5, for the sake of completeness a brief review of methods 1, 2 and 3 is also given.

1. Mechanical

Mechanical means of removing scale is still practiced. One method other than chipping and drilling is to use thermal shock. This is only effective in submerged tube evaporators where the sea-water brine being evaporated is contained in a pool in which steam-heating surfaces are immersed. If, when the system is cold, a charge of live steam is sent through the heating surfaces, the resulting thermal shock is often sufficient to loosen the deposited scale. This is often done in conjunction with chelating agents such as derivatives of sulphonic acid which weaken the bond between scale and heating surface. This method is commonly used on ship-board evaporators.

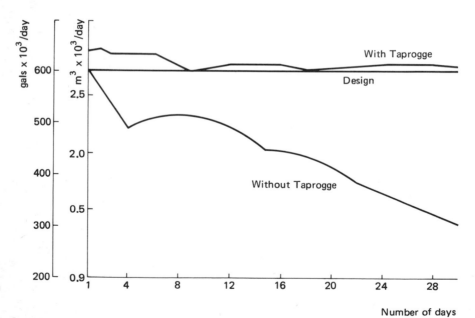

Fig. 3.3. Output variation for a polyphosphate-dosed plant with and without Taprogge installation.

One novel mechanical method of removing soft sludges if deposited inside tubes (often encountered with treatment in method (4)) is the 'Taprogge' system described by Elliott [6] and Querns.[7] This consists of passing flexible rubber balls through the heater tubes, resulting in a scouring action which effectively removes the sludge accumulation. The results of using the Taprogge system are worth mentioning in detail as the method is commonly used on polyphosphate-dosed plants as in the 2 x 4 546 m^3/day (2 x m.g.d.) MSF plants in Qatar and 3 x 4 546 m^3/day (3 x m.g.d.) MSF plants in Malta. [7] Figure 3.3 shows the performance maintenance which is possible when the Taprogge system is used in conjunction with phosphate dosing discussed in sub-section 4.

2. Seeding

In any commercial plant nucleation sites for scale formation are readily available. If, however, preferential nucleation sites are provided and if these sites can be removed, processed and returned to the system economically, then a cheap method of scale prevention is available. This method is known as seeding, whereby a finely dispersed substance (seeds) corresponding to the type of scale which is to be removed is introduced into the sea-water brine stream. When supersaturation occurs, preferential deposition takes place on the seeds if sufficient residence time is given in the supersaturated solution. As the removal and subsequent grinding down of the used seeds to optimum size is expensive, this method is not extensively employed for the prevention of alkaline scales.

The use of seeding in a multiple effect plant has been reported in the USSR [8] and in a vapour compression plant in the US. [9] It cannot, however, be considered a proven process but it has potential for the prevention of calcium sulphate scale which, if it is cost-effective will permit higher operating temperatures.

3. Ion exchange

Ion exchange methods consist of passing the raw feed through a resin bed which replaces calcium and magnesium ions with sodium ions as given in equation (3.5)

$$R - Na + \tfrac{1}{2}Ca^{++} \rightarrow R - \tfrac{1}{2}Ca + Na^+ \tag{3.5}$$

The resin eventually becomes spent and requires regeneration with a concentrated brine stream as given in equation (3.6)

$$R - \tfrac{1}{2}Ca + Na^+ \rightarrow R - Na + \tfrac{1}{2}Ca^{++} \tag{3.6}$$

This method is termed base exchange and is often used for treating boiler feed-water but is extremely expensive to use on sea water because of the high concentrations of scale-forming salts. This necessitates high resin bed capacities and frequent regeneration with a concentrated caustic soda brine stream as attempts to use reject sea-water brine have so far not been successful. Another factor which militates against this process is the break-down or elutriation of the resin bed which means resin replacement costs are high.

4. Proprietary compounds

Scale inhibition with proprietary compounds is a commonly used method and is termed threshold treatment, as the quantity employed is much less than the stoichiometric amounts required to combine with the calcium and magnesium present in the feed. The principal proprietary compound is Hagevap which is a mixture of sodium polyphosphates, lignin sulphonic acid derivatives and various esters of polyalkylene glycols. The dosing quantities employed range from 2–5 p.p.m. which are effective up to temperatures of around 85°C (185°F). The mechanism by which Hagevap prevents scale deposition is not clearly understood. One explanation is that the polyphosphate chain enters the scale crystal lattice while it is at an embryo stage and prevents further bond formation which results in a finely dispersed micro-sludge.

As the temperature of the sea-water brine is increased, the polyphosphate molecules become hydrolysed and then combine with magnesium ions to form magnesium phosphate which deposits as a sludge on the heat exchange surface. This hydrolysis reaction places a temperature limitation on the use of polyphosphate compounds to less than 85°C (185°F). The use of Hagevap at a maximum temperature of 93°C (200°F) and a concentration factor of 1.5 in a 4 546 m³/day (1 m.g.d.) MSF plant is reported by Mulford *et al.* [10] who observed the occurrence of light sludge deposits as evidenced by a rise in pressure and temperature of the steam supply to the brine heater and a fall in the plant

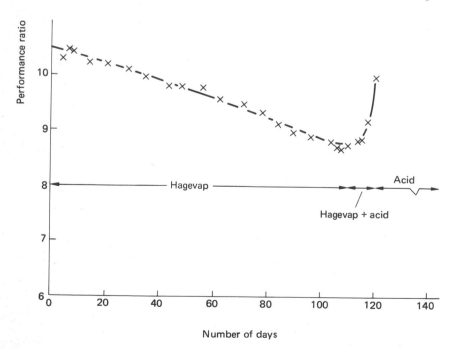

Fig. 3.4. Performance ratio variation on a polyphosphate-dosed plant.

performance ratio as shown in Fig. 3.4 from reference. [10] Any sludges which do form as a result of polyphosphate hydrolysis may be readily removed by acid slugging, i.e. a pulse of dilute acid is sent round the plant at intervals dictated by the reduction in plant performance. The rate of deposition may also be controlled by the water side velocity in MSF plants where a velocity of order 2 m/s (6.5 f/s) can maintain a limiting sludge deposit.

The design penalties of using polyphosphate treatment is a reduction in the overall heat transfer coefficient which in MSF plants can be of the order of 567 W/m^2 °C/week (100 Btu/h ft^2 °F/week) initially. However, if enough heat transfer surface is built into the plant, an equilibrium situation is reached and acid cleaning may only be required at roughly six-monthly intervals. Reported polyphosphate treatment costs for the state of Kuwait are approximately 2 per cent of product cost. [11] The use of the Taprogge system for maintaining heat transfer surface performance has already been mentioned and as reference to Fig. 3.3 shows, this method can reliably maintain plant output.

The use of threshold treatment for the prevention of $Mg(OH)_2$ scale in the 85°–107°C (185°–225°F) range is reported by Elliot and Hodgson. [33] The principles are similar to the polyphosphate treatment described previously, viz. a delay in the onset of precipitation and a deformation of the crystal lattice so that the scale does not adhere. The distorted crystals form a soft porous deposit in contrast to the hard crystalline $Mg(OH)_2$ scale. This deposit has a high heat transfer resistance but is very easily removed. Related work on theoretical mechanisms of magnesium scale formation has been reported by Harris and Finan. [34] The successful pilot plant tests using this new additive for $Mg(OH)_2$ prevention mean that the flash range can be extended before acid treatment has to be adopted.

5. pH control

Acid treatment is a very successful method of preventing alkaline scale formation and consists of supplying hydrogen ions to break down the bicarbonate ions. Thus:

$$HCO_3- + H^+ \rightleftharpoons CO_2^\uparrow + H_2O \qquad (3.7)$$

The source of supply of hydrogen ions is dictated by the cost of the stoichiometric amount required to treat the bicarbonate content of the feed. On this basis sulphuric acid is usually the cheapest source of hydrogen ion followed by hydrochloric acid and ferric chloride which is often used in shipboard evaporators.

As an example of typical feed treating costs, consider the use of sulphuric acid 95 per cent concentration which is available at 2.2 pence/kg (1 pence/lb) for the treatment of a raw sea-water feed with 120 p.p.m. of alkalinity expressed as $CaCO_3$. The stoichiometric amount of sulphuric acid required follows from equation (3.8) below.

$$CaCO_3 + H_2SO_4 \rightarrow CaSO_4 + H_2O + CO_2 \qquad (3.8)$$

i.e. 1 mole sulphuric acid (98 lb) is required for 1 mole of $CaCO_3$ (100 lb). For 95 per cent concentrated acid, the dosing rate would be 98/95 × 120 = 124 p.p.m.

Assuming that the plant concentration factor is 2, then the raw feed-water for 1 000 gallons of product is 2 000 gallons, thus the unit cost of treating the feed with sulphuric acid is

$$2\ 000 \times 10 \times \frac{124}{10^6} \times 1 = 2.48 \text{ p/1 000 gal product}$$

$$(\text{or } 2 \times 1\ 000 \times \frac{124}{106} \times 2.2 = 0.545 \text{ p/m}^3)$$

Raw feed treatment costs can be estimated in a similar manner for any other combination of acid cost and plant feed parameters. In areas where acid is expensive such as Kuwait, acid dosing could add 1.21 p/m³ (5.5 p/1 000 gallons) to the product cost.

Precautions must be taken with acid dosing to prevent the circulation of dilute acid solutions which are highly corrosive. Two common methods of preventing corrosion with acid dosing are:

(a) The combined use of a CO_2 degasser followed by the addition of a small amount of caustic soda to raise the feed pH value to around 8.
(b) The other method is to use slightly less than stoichiometric amounts of acid and thus leave some residual alkalinity in the untreated feed which results in a pH around 7.7.

In both cases CO_2 degassing is required which may be by cooling-tower methods using a packed column. The acid-dosed feed is passed downwards through a packed column in counter current contact with a flow of air which sweeps the CO_2 from the feed which is already saturated with oxygen. The CO_2 degassed feed is then sent to a vacuum deaerator where the residual CO_2 and air are reduced to a level deemed sufficient to eliminate corrosion.

The choice of feed treatment should be an economic one especially for large desalination plants, i.e. the feed treatment choice is determined on the basis of total product cost resulting from the method used. It is merely noted just now that the fraction of total product cost due to feed treatment is not the materials cost alone but also the cost of including design changes such as excess heat transfer surface to allow for the fouling which accompanies polyphosphate dosing. More will be said on this subject after calcium sulphate scale formation has been discussed.

Calcium sulphate scaling

Calcium sulphate scaling occurs as a result of its reverse solubility in sea water, i.e. as sea-water brine temperatures and concentrations increase, the $CaSO_4$ solubility in sea water decreases. As calcium sulphate is not at saturation level in normal sea water, it is possible to prevent this scale forming if the distillation process is operated below temperatures and concentrations at which it occurs. As this scale cannot be readily removed by acid cleaning, its deposition must be prevented.

Figure 3.5 shows the theoretical scaling limits of calcium sulphate. It is seen that three forms of deposition can occur, each being operative over a given temperature range. These three forms and their chemical compositions are:

(i) gypsum, $CaSO_4 . 2H_2O$
(ii) hemi-hydrate, $CaSO_4 . \frac{1}{2}H_2O$
(iii) anhydrite, $CaSO_4$

Each form has its respective solubility range, the anhydrite being the least soluble and therefore the one which would be deposited first. However, for scaling to occur, three requirements must be met. These are, availability of nucleation sites, supersaturation of the scaling compound, adequate residence or nucleation time.
    The first and third conditions are readily fulfilled in most distillation plants. Thus the prevention of calcium sulphate scaling is done by the non-fulfilment of the second condition with the added proviso that the anhydrite form requires a long residence time for its occurrence. It is thus possible to run a distillation plant to the right of the anhydrite line in Fig. 3.5, but to the left of the hemi-hydrate solubility line. This permits higher operating temperatures at a given brine concentration. This method is commonly practiced in acid-dosed MSF plants as the flashing brine spends a relatively short time at the high temperature end of the plant. Should any calcium sulphate scale nuclei be formed, they either redissolve or

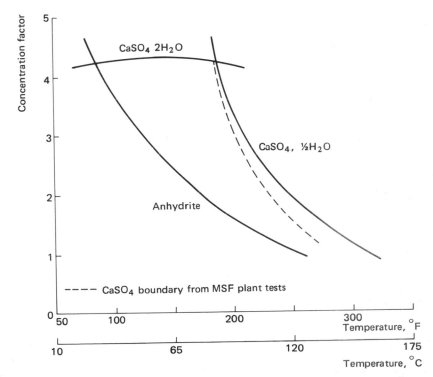

Fig. 3.5. Solubility of $CaSO_4$ for varying sea-water concentration and temperature.

do not crystallise when the brine is cooled below the point of supersaturation when it reaches the low temperature end of the plant.

It should be noted that in the boundary layer adjacent to the tube wall, the retention time may be such that anhydrite scaling can occur. This is governed by the combination of residence time at the wall and the wall temperature which at

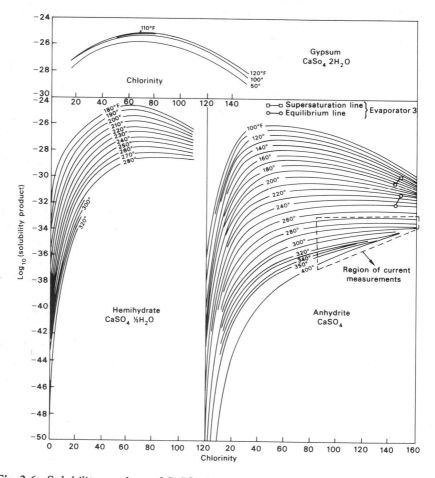

Fig. 3.6. Solubility produce of $CaSO_4$ in sea-water concentrate.

Solubility product....[Ca] [SO_4] in $(g.mols)^2 (kg)^{-2}$
Chlorinity in.............g./1000g.

high heat flux operation can be several degrees warmer than the bulk fluid temperature as discussed in [12]. Generally, operation above the anhydrite solubility limits is possible and avoidance of the gypsum and hemi-hydrate solubility limits is required. Much experimental work has been done in this field [3–5, 13, 14] and a useful result is Fig. 3.6 from the work of Simpson and Hutchinson [14] which shows the variation in solubility product with chlorinity (or concentration) of the sea-water brine with temperature as a parameter. The solubility product K is defined as the product of the molar concentration of ions present, i.e.

$$K \: CaSO_4 = [Ca^{++}] . [SO_4^{--}]$$

at fixed temperature while the [ ] symbol denotes molar concentration.

As the solubility of a compound is governed by the solubility product, it is a relatively easy matter to calculate the maximum temperature to which sea-water brine may be heated for any given concentration. Thus for a concentration factor of 2, i.e. a chlorinity value of 40 g/1 000 g and from Table 3.2

$$[Ca^{++}] = 2 . (0.010 \: 5) \: g \: mol/l$$

$$[CO_4^{--}] = 2 . (0.028 \: 8) \: g \: mol/l$$

Thus the product of the molar ionic concentrations is

$$[Ca^{++}] . [CO_4^{--}] = 1.2 \times 10^{-3} = - \log_{10} 2.9$$

Referring to Fig. 3.6 on the hemi-hydrate curves at a chlorinity of 40 and $\log_{10}$ solubility product of -2.9 gives a maximum temperature of $126°C$ ($260°F$) to which sea-water brine of concentration factor 2 can be heated without risk of hemi-hydrate deposition.

At the other extreme, sea water of concentration factor 1, i.e. $\log_{10}$ solubility product = $\log_{10} 3.523$ can be heated to a theoretical maximum temperature of $160°C$ ($320°F$) without risk of hemi-hydrate formation. This theroretical maximum must be treated with caution as actual MSF plant maximum operating temperatures at a concentration factor of 1 have been nearer $143°C$ ($290°F$). [15]

It is seen that calcium sulphate scaling considerations currently fixes the top temperature at which distillation plants can operate at any fixed concentration factor. The incentives for increasing operating temperature may be readily grasped from the following simplified example.

Consider an MSF plant using Hagevap feed treatment with a flash range of $50°C$ ($90°F$), i.e. a top temperature of $88°C$ ($190°F$) which uses steam near atmospheric pressure (10^5 Pa) in the brine heater; and another MSF plant using an acid-dosed feed with a flash range of $89°C$ ($160°F$), i.e. top temperature $127°C$ ($260°F$), bottom temperature $38°C$ ($100°F$) using steam near 3.10^5 Pa (30 p.s.i.g.) in the brine heater. Then, all other things being equal, the $89°C$ ($160°F$) flash range plant will produce approximately 1.75 times the output of the $50°C$ ($90°F$) flash range plant, as output is proportional to flash range if the brine recirculation quantities on each plant are equal.

The latter plant gives a much greater output than the former for much the same energy input though the 3.10^5 Pa (30 p.s.i.g.) steam may be more expensive to purchase than the atmospheric pressure steam. There are clear incentives for increasing the top temperatures in MSF plants as generally the smaller the flash range the greater is the capital cost for a given output and much research is being devoted to increasing the top temperature limit. One proposal is the multi-effect multi-stage plant (MEMS) which uses brine with a concentration factor near unity in a high temperature flashing effect consisting of several flashing stages before cascading the brine to another lower temperature effect where it acts as the feed for a brine stream of greater concentration factor.

In this way the flashing range may be considerably extended albeit at the price of plant complexity. A proposal to operate a single-effect multi-stage (SEMS) plant at a top temperature of 149°C (300°F) has also been reported. [16]

Feed treatment selection

The choice of feed treatment is determined mainly by its contribution to product water cost including the effects of design changes to accommodate the particular method chosen. On the one hand acid treatment allows higher operating temperatures up to 127°C (260°F) in conventional MSF plants. On the other, polyphosphate dosing restricts top temperatures to roughly 89°C (190°F) and the designer may be called upon to insert 23 per cent extra heat transfer surface compared with the equivalent acid-dosed plant. However, acid dosing can be more expensive than polyphosphate dosing.

In areas where acid costs are expensive, e.g. Kuwait where acid can cost as much as £45 per tonne, acid dosing would add 1.03 p/m^3 (4.7 p/1 000 gal) of product whereas polyphosphate dosing at £200 per tonne and a dosing rate of 2$\frac{1}{2}$ p.p.m. of feed would add only 0.1 p/m^3 (0.46 p/1 000 gal) of product. In this case the choice is clear-cut and polyphosphate dosing is chosen despite the added costs of increased heat transfer surface and lower flash range resulting in a costlier plant for a given output. However, for areas such as the US where acid costs are approximately £10 per tonne the dosing cost can approach 1 p/1 000 gal product. In this case acid dosing would be chosen given that the cost of the higher pressure steam required for the brine heater would not render this choice uneconomic.

Other feed treatment methods such as the Grace precipitation method [17] are often proposed to enable plants to run at much higher top temperatures than those dictated by calcium sulphate scaling. The economic practicality of these methods has yet to be proven. In addition the steam cost at the brine heater will rise quite steeply as its thermodynamic availability increases. Thus an economic constraint may be operative whereby top temperatures in excess of 149°C (300°F) are not worth pursuing.

Corrosion/erosion processes

The effects of corrosion alone or in combination with erosion are a constant hazard in sea-water distillation plants. Two types of conditions are encountered – cold aerated sea water and hot 'deaerated' sea-water brine. Each condition requires special provision for the prevention and control of corrosion. Before discussing these methods, the corrosion processes and the accompanying effects which may be encountered in distillation plants are briefly discussed. The respective texts by

Speller [18] and Uhlig [19] give a thorough and definitive treatment of the corrosion field.

General corrosion

This is approximately uniform metal removal not accompanied by localised action such as pitting, cracking or erosion. The cause is electrochemical in nature. All industrial metals have heterogeneities of some form such as slag inclusions, mill scale, and localised stresses. These inhomogeneities form local anodic and cathodic areas within the same piece of metal, consequently metal ions are removed at the anode when the metal is immersed in an electrolyte. The metal heterogeneities and the accompanying electrochemical corrosion cause a constant shifting of the anodic and cathodic areas in the exposed surfaces so that uniform corrosion results. The corrosion of ferrous compounds such as iron and steel is a pertinent example. The presence of an aqueous solution and oxygen is essential and the presence of carbon dioxide and metallic salts will accelerate the process. The inhomogeneities in the wetted metal surface form a series of short circuited electrochemical cells. At the anode atoms of metallic iron lose two electrons each and enter the solution as Fe^{++}.

At the cathode, water provides both hydrogen and hydroxyl ions, H^+ and OH^- respectively. The electrons from the anode combine with the hydrogen ions to yield hydrogen gas, thus $2e + 2H^+ = H_2$. In the presence of oxygen the hydrogen is oxidised to water and the metallic ions combine with the hydroxyl ions. Thus $Fe^{++} + 2OH^- = Fe(OH)_2$ ferric hydroxide which in the presence of carbon dioxide, hydrolyses to form hydrated ferric oxide or rust. The rate of corrosion of steel is governed by the diffusion of oxygen to the metal surface, thus moving aerated sea water is an ideal corrosive medium as it supplies fresh oxygen to the cathode. In practice general corrosion can be tolerated as corrosion rates are of the order 0.012 cm (0.005 in) penetration per year, in aerated sea water. Hence a corrosion allowance of 1 cm is a very adequate general corrosion allowance for a 25-year life in brine vessels operating at temperatures below $200°F$.

Galvanic corrosion

When two dissimilar metals are in contact in an electrolyte the less noble, i.e. anodic, of the two corrodes because of the flow of electrons resulting from the electropotential difference between the two metals. The following metals form a galvanic series in sea water. [20]

Galvanised wrought iron
Mild steel
Cast iron
Ni-resist
Manganese bronze

Admiralty brass
Copper
70/30 Cupro nickel
18/8 Stainless steel
Nickel.

The most anodic is first, i.e. the metal which is joined with any succeeding metal in the series will form the anode and corrode. The severity of galvanic corrosion depends on both the difference in electropotential between the materials and the respective surface areas involved. If the area of the cathodic metal is large compared with that of the anodic metal, the severity of the corrosion is much greater. These electrolytic effects are often severe in water boxes where there are large concentrations of non-ferrous surfaces such as the heat exchanger tubing. Where this is so a sacrificial anode may be used to protect the tube plates.

Localised corrosion

Pitting is one of the most severe forms of localised corrosion. It can occur very rapidly due to the local failure of the protective oxide film which forms on certain metals such as stainless steels. When the protective film is broken a corrosion cell is set up which does not move and rapid corrosion occurs at the anode. This form of corrosion is common in aluminium and stainless steel alloys in the presence of metal chlorides. Hence it can be extremely severe in a sea-water environment.

Impingement and cavitation corrosion

Impingement or erosion corrosion results when a moving fluid removes protective films from metal surfaces faster than they can reform. Thus in areas where there are flow obstructions and high eddy velocities such as the tube inlets of a water box, this form of attack may be expected. In pumps cavitation can occur, i.e. swarms of vapour bubbles are formed and collapse with an associated impact on metal surfaces. The result of this continual impacting can be spectacular failure in the components involved.

Corrosion caused by polluted sea water

Various other forms of corrosion can be encountered in desalination plants. One example is copper attack which may be coupled with intergranular corrosion where a polluted sea-water feed is used which may contain sewage. Decomposition of the contaminants reduces the oxygen content of the water and gives rise to sulphide and ammonium compounds in solution.

The reduction in oxygen content severely reduces the film-forming properties of the tube materials and the sulphides and ammonium compounds give rise to an aggressive feed. This type of feed attacks the copper base alloys and in the case of aluminium brass severe intergranular corrosion can occur where the grain boundaries of the alloy are anodic to the grain bulk. This form of corrosion is discussed in Todd [21] and Uhlig. [22]

Bacterial corrosion

In the metabolism of any living organism, there is energy intake and an energy release. In some bacteria these energy changes result in a measurable electric current. When corrosion-causing organisms reproduce on the surface of an iron pipe, electrobiochemical removal of iron takes place.

Micro-organisms then, can participate in iron and steel corrosion by creating favourable conditions for electrochemical reaction to take place. One mechanism is the changing of a surface film resistance with metabolic products such as sulphuric or organic acids. In other instances slime deposits can form in selected areas so that they become anaerobic. If oxygen is present in solution the anaerobic sites become anodic to the aerobic areas.

When two portions of the same metal receive oxygen at different rates, a corrosion cell is established and two types of oxidation can take place – dehydrogenation and loss of electrons.

The hydrogen atom dissociates into a proton and an electron. Electrons are transported by an electron-carrying system and pass through the corroding metal to the cathodes; where each electron displaces a positive ion, usually hydrogen. The positive metal ions formed at the anodes go into solution.

Sulphate reducers

The sulphate reducers are the most important corrosion-causing bacteria. They use the sulphate ion as the terminal electron acceptor for respiration. They obtain energy by the reduction of the sulphate ion by hydrogen which comes from molecular hydrogen or organic compounds. By virtue of their metabolism the organisms use molecular hydrogen and produce H_2S. Corrosion is by cathodic depolarisation and is increased by the organisms' consumption of hydrogen.

Equations (3.9) to (3.13) below illustrate the chemical reactions which take place.

Anodic

$$8H_2O \rightleftharpoons 8H^+ + 8(OH)^- \tag{3.9}$$

$$4Fe + 8H^+ \rightleftharpoons 4Fe^{++}8H \tag{3.10}$$

Cathodic depolarisation (bacteria)

$$SO_4^= + 12H \rightleftharpoons H_2S + 2H_2 + 2H_2O + 2(OH)^- \tag{3.11}$$

Corrosion products

$$Fe^{++} + H_2S \rightleftharpoons FeS + 2H^+ \tag{3.12}$$

$$3Fe^{+++} + 9(OH)^- \rightleftharpoons 3Fe(OH)_3 \tag{3.13}$$

As equations (3.9) to (3.13) show, the result is the continuous removal of iron from the anodic areas and pitting takes place. FeS is cathodic to metallic iron and its presence accelerates corrosion.

Other bacteria can oxidise H_2S for their metabolic energy supply as shown in equations (3.14) and (3.15).

$$2H_2S + O_2 \longrightarrow 2S + 2H_2O \tag{3.14}$$

$$2S + 3O_2 + 2H_2O \longrightarrow 2H_2SO_4 \tag{3.15}$$

The H_2SO_4 produced is highly corrosive and forms an anodic concentration cell that further accelerates corrosion (a pH of 0.2 has been measured).

Another class of bacteria commonly termed 'iron bacteria' converts ferrous oxide to ferric hydroxide for their energy source. The ferric hydroxide covers the organisms which deposit on pipe walls creating areas of ferric hydroxide. The anaerobic conditions created provide ideal growth conditions for sulphate-reducing bacteria. The continued production of iron enables additional iron bacteria to grow on the deposits. The cycle is repeated and severe pitting may result.

Bacterial action can be stopped by eliminating the conditions suitable for their growth. This, however, is often impossible if, for example, their nutrients come in the feed-water. Resort must be made to toxicants to control the organisms, eradication being impossible if the feed contains their food supply. Usually chlorine dosing of the feed is a sufficient treatment. Woods [23] lists various biocides and discusses their effectiveness.

Corrosion prevention

From the foregoing discussion it is obvious that many factors conducive to corrosion/corrosion–erosion are present in desalination plants. The methods which are used to prevent or reduce its occurrence may be summarised as:

1. Removal of corrosive agents in the feed
2. Avoidance of electrolytic cells
3. Protective coatings
4. Proper materials selection.

1. Removal of corrosive agents

The two main corrosive components in the sea-water feed are dissolved oxygen and carbon dioxide. Removal can be either of two ways. One method is to use a vacuum desorption tower where the feed is sprayed in at the top and allowed to partially flash. This removes most of the oxygen and carbon dioxide along with some vapour. This type of unit is expensive as the materials used must be corrosion resistant or heavily protected.

A more common approach is to separate the oxygen and CO_2 desorption duties. If the feed is acid-treated it will be rich in CO_2 and is fed to an atmospheric pressure stripping tower. This consists of a packed column with the feed trickling downwards in counter-current contact with a stream of air drawn upwards by induced draught fans. The CO_2 is virtually completely removed by this method and for an evaporator plant the feed is sent to the heat rejection section which operates at very low pressures and is injected into an appropriate high vacuum stage for deaeration in the stage itself. A separate deaerator can of course be used for installations where this method is inappropriate.

With the CO_2 and dissolved oxygen removed, the feed corrosive properties are greatly reduced and normally it can be circulated around the plant without further treatment. The pH should be between 7.5 and 8 to prevent the circulation of dilute acid solutions.

2. Avoidance of electrolytic cells

Electrolytic effects predominate in water boxes where there are large non-ferrous concentrations such as tube banks and tube sheets. In cases where the water boxes are rubber-lined soft iron sacrificial anodes are fitted to produce iron in solution for the protection of the non-ferrous tubes. [24] The use of impressed current protection for consider tubes in general is discussed in reference. [25]

3. Protective coatings

This is a common method of mitigating the effects of corrosion. In brine vessels operating above 93°C (200°F) common practice is to use copper-clad steel sheeting. Where the sea water is below 49°C (120°F) and aerated, feed lines and brine vessels are often cement-lined carbon steel or tar mastic to prevent oxygen access to the steel surface. Rubber linings are often used in water boxes. Where corrosion-erosion is expected to be more severe, such as ends of heat exchanger tubes, plastic inserts are occasionally used [26] as shown in Fig. 3.7. More information in protective coatings is given in Table 3.4.

4. Materials selection

This is obviously of the utmost importance and is best illustrated in tubular form. Two tables are given. Table 3.3 gives the relative cost ranking for 2.54 cm (1 in) diameter 0.122 cm thick (18 s.w.g.) non-ferrous tubing subject to hot high velocity sea water and comments on its applicability in various corrosion–erosion situations. Attention has been focused on the heat exchanger tubing as 35–45 per cent of most large distillation plant capital costs are attributable to the heat transfer surface.

Fig. 3.7. Tube inlet-end protection by plastic insert.

Table 3.3. Ranking of corrosion-erosion properties of non-ferrous metals in hot sea water

Material	Relative cost [28]	Maximum velocity for 20-year life where available	Comments as to suitability
Titanium 0.091 cm (20 s.w.g. thickness)	3.6	Has been used at 2.45 m/sec (8 ft/sec) at 99°C (210°F)	Used in US plant in Virgin Islands. Good wetability. Extremely good corrosion resistance. Max. temp. 126°C (260°F).
Stainless steel 316	1.4	Greater than 3 m/sec (10 ft/sec)	Extremely good resistance to CO_2 environment, e.g. condenser tubing for air ejectors. Resistant to acidic condensates, susceptible to pitting on brine side. Suitable for above 93°C (200°F)
70/30 cupro nickel	1.0	4.5 m/sec (15 ft/sec)	Good erosion resistance
90/10 cupro nickel	0.9	3.65 m/sec (12 ft/sec)	More resistant to fouling by marine organisms [30] than 70/30 or al-brass. Competitive with al-brass in 0.091 cm (20 s.w.g.) thickness. [21] A very good material cost-wise.
Aluminium brass	0.6	2.1 m/sec (7 ft/sec)	Intergrannular corrosion can occur in polluted sea water [21]. Complete tube failures recorded in Virgin Islands
Mild steel	0.15		Failed after 2 years due to corrosion on vapour side:[31] may be suitable for use below 60°C (140°F)[27] in deaerated brine.

Table 3.4. Materials summary – A. Aerated sea water 49°C (120°F) max.

Vessels	1. Steel coated with coal-tar mastic or cathodically protected. 2. Cast-iron coated with coal-tar mastic or cathodically protected.
Pumps	1. 316 SS 2. Ni-Resist Type 1B or Type D-2
Lines	1. Cement-lined carbon steel 2. Asbestos-cement – max. 1 m (3.2 ft) diameter 3. Glass fibre reinforced plastic pipe – max. 0.5 m (21 in) diameter 4. Polyvinyl chloride – max. 0.3 m (12 in) diameter
Valves	1. Rubber-lined butterfly 2. Iron body, brass trim 3. Brass 4. Cupro-nickel 5. Ni-Resist
Expansion bellows	1. Monel 2. 316 SS 3. Cupro-nickel
Structures	1. Steel coated or cathodically protected
Screens	1. Monel 2. Brass 3. Cupro-nickel
Tubeplates	1. Carbon steel, rubber coated 2. Cast iron 3. Cupro-nickel

Table 3.4 summarises the principal items of equipment found in evaporation plants and the range of duties adjudged suitable. This table is based on the materials review by White. [27] A good review of coastal power station condenser practice, which has many problems in common with distillation processes, is given by Pexton. [25]

So far most distillation plants have been built with the cheapest available materials consistent with 'reasonable' life expectancy. This has led to carbon steel chambers which have later had to be replaced by stainless steel cladding as was done at the Key West installation (2.2 m.g.d.). Tubing made from aluminium brass was replaced with 90/10 Cu Ni (St. Thomas). Briefly, carbon steel alone should not be used at temperatures above 93°C (200°F). The 'Clare Engel' plant in the US has had some carbon steel piping replaced twice in three years. [32]

The practice of using low-cost materials with their attendant corrosion problems has dominated distillation plant design. The alternative route of avoiding corrosion by the use of more costly materials is not popular but a body of opinion is now behind this approach. Serious corrosion of low carbon steel in sea-water brine

Table 3.4. Materials summary – B. Deaerated sea-water brine.

	Max. 65°C (150°F)	65°-93°C (150°-200°F)	93°-120°C (200°-250°F)	120°-159°C (250°-300°F)
Vessels	1. Concrete lined with asphaltic mastic	Concrete lined with hot cured neoprene mastic	Concrete lined with neoprene sheet	
	2. Carbon steel + 0.65 (1/4 in) corrosion allowance	Carbon steel clad with stainless steel	Carbon steel clad with cupro-nickel or coated	
	3. Carbon steel clad with epoxy			
	4. Cast iron		Coated cast iron	
	5. Ni-Resist Type 1B(1) or Type D-2			
Pumps	1. Carbon steel			
	2. 316 SS			
	3. Ni-Resist type 1B or Type D-2			
Lines	1. Carbon steel	Carbon steel + 0.65 cm (1/4 in) corrosion allowance	Coated or cement-lined carbon steel	
	2. Coated or cement-lined carbon steel	Stainless steel	Stainless steel clad with carbon steel	
Elbows	3. Stainless steel			
Expansion bellows	1. Monel			
	2. 316 SS			
	3. Cupro-nickel			
Troughs	1. Coated carbon steel			
	2. Aluminium brass			
	3. Cupro-nickel			
Demisters	1. Monel Liable to hydrogen-sulphide attack			
	2. Cupro-nickel			
	3. Stainless steel			
Tubeplates	1. Carbon steel rubber coated		Clad carbon steel	
	2. Cupro-nickel			

temperatures above 93°C (200°F) has been found in many plants. Protective coatings also appear to be of little use and one can expect future plants to have much more attention paid to materials selection in view of the accumulation of operating experience now at hand.

References

1. Grunberg, L. Proc. 1st Int. Symp. on Water Desalination, Washington D.C., Oct. 1965, p. 157.
2. Emmerson, W. H. and Jamieson, T. D. *Desalination,* **3** 1967, p. 213.
3. Hillier, H. *Proc. I. Mech. E.,* **1B**, 1952, p. 295.
4. Standiford, F. C. and Sinek, J. R. *Chem. Eng. Prog.,* **57**, 1961, p. 58.
5. Langelier, W. F., Caldwell, D. H. and Lawrence, W. B. *Ind. Eng. Chem.,* **42**, 1950a, p. 126.
6. Elliott, M. N. U.K.A.E.R.E. Report R–5820, Sept. 1968.
7. Querns, W. Paper SM–*113/43*, I.A.E.A. Symp. on Nuclear Desalination, Madrid, Nov. 1968.
8. Cherrozubov, V. B. *et al.* Proc. 1st Int. Symp. on Water Desalination, Washington D.C., Oct. 1965, p. 539.
9. Geirenger, D. L. Proc. 1st Int. Symp. on Water Desalination, Washington D.C., Oct. 1965, p. 659.
10. Mulford, S., Glater, J. and McCatchan, J. W. 1st Int. Symp. on Water Desalination, Washington D.C., Oct. 1965, p. 139.
11. Ali el Saie, M. H. Proc. 1st Int. Symp. on Water Desalination, Washington D.C., Oct. 1965, p. 287.
12. Tidball, R. A. and Woodbury, R. E. Proc. 1st Int. Symp. on Water Desalination, Washington D.C., Oct. 1965, p. 317.
13. Flint, O. and Elliot, M. N. Paper 37, 2nd European Symp. on Fresh Water from the Sea, Athens, May 1967.
14. Simpson, H. C. and Hutchinson, M. Paper 43, 2nd European Symp. on Fresh Water from the Sea, Athens, May 1967.
15. Baldwin-Lima-Hamilton Corp. US Office of Saline Water Progress Report 288, Oct. 1967.
16. Tidball, R. A. and Gaydon, J. G. Paper SM–113/26, I.A.E.A. Symp. on Nuclear Desalination, Madrid, Nov. 1968.
17. W. R. Grace & Co. US Office of Saline Water. Research and Development Report No. 192, May 1966.
18. Speller, N. *Corrosion – Causes and Prevention*, McGraw-Hill, New York, 1951.
19. Uhlig, H. H. (ed.). *Corrosion Handbook*, J. Wiley and Sons Inc., New York, 1948.
20. Todd, B. 'The Corrosion of Metals in Desalination Plants', *Desalination,* **3**, 1967.
21. Todd, B. Paper 49, 2nd European Symp. on Fresh Water from the Sea, Athens, May 1967.
22. Uhlig, H. H. Proc. 1st Int. Symp. on Water Desalination, Washington, D.C., Oct. 1965, p. 171.

23. Woods, G. A. 'Bacteria friends or foes?' *Chem. Eng.,* 5 March 1973, 81–4.
24. Stewart, J. M. Proc. 1st Int. Symp. on Water Desalination, Washington, D.C., Oct. 1965, p. 651.
25. Pexton, A. F. 'The Influence of Condensers on Modern Power Station Design', Paper presented at Symposium James Watt Bicentenary, University of Glasgow, Sept. 1969.
26. Van der Bergh *et al.* Proc. 1st Int. Symp. on Water Desalination, Washington, C.C., p. 189.
27. White, R. A. *Materials Protection,* March 1965, p. 48.
28. Vilentchuk, I. *Desalination,* **4**, 1968.
29. Feige, N. G. Paper SM-113/25, I.A.E.A. Symp. on Nuclear Desalination, Madrid, Nov. 1968.
30. Noguchi, N. *et al.* Paper SM-113/12, I.A.E.A. Symp. on Nuclear Desalination, Madrid, Nov. 1968.
31. Foster, A. C. and Herlihy, J. P. Proc. 1st Int. Symp. on Water Desalination, Washington, D.C., Oct. 1965, p. 57.
32. Evans, R. H. and Mulford, S. F. Proc. 3rd Int. Symp. on Fresh Water from the Sea, Vol. 4, p. 43, Djbrovnik, 1970.
33. Elliot, M. N. and Hodgson, T. D. 'Additives for alkaline scale control at temperatures above 93°C (200°F)', Proc. 4th Int. Symp. on Fresh Water from the Sea, Vol. 2, pp. 97–110, Heidelberg, 1973.
34. Harris, A. and Finan, M. A. 'A theory of the formation of magnesium scales in sea water distillation plants and means for their prevention', Proc. 4th Int. Symp. on Fresh Water from the Sea, Vol. 2, pp. 131–42, Heidelberg, 1973.

Chapter 4

Multi-stage flash distillation

Introduction

Multi-stage flash (MSF) distillation, currently the dominant saline water conversion process, has almost entirely superseded its predecessor – submerged tube boiling distillation. In fact as Silver [1] points out, for submerged tube boiling distillation an average rate of growth was 10 000 m³/day/year from 1952–60 and 700 m³/day/year thereafter since the MSF process evolved. For multi-stage flash distillation the growth rate is 25 000 m³/day/year from 1960 onwards. Thus MSF has been responsible for a dramatic increase in installed capacity of land-based distillation plants – so much so that due consideration of other processes such as the vertical tube evaporator (VTE) is sometimes overshadowed by its numerical superiority.

The likely market in terms of installed sea water conversion capacity from its current 1.8×10^6 m³/day (400 m.g.d.) has been given by Pugh [2] as a total market of 7.2×10^6 m³/day (1 600 m.g.d.) by 1990 with the following confidence limits

Per cent		m^3/day	(m.g.d.)
90	confidence in an additional	1.6×10^6	360
75	confidence in an additional	2.72×10^6	600
50	confidence in an additional	2.27×10^6	500

This growth is estimated to come from countries with a lack of access to alternative supplies. Pugh also states that of those countries with access to the sea:

(a) Five to six countries are expected to install over 0.725×10^6/day (160 m.g.d.) this century;

(b) Eleven to twelve countries are likely to need supplies between 4.54×10^4–4.54×10^5 m³/day (10–100 m.g.d.);

(c) Sixteen countries need between 4.54×10^3–4.54×10^4 m³/day (1–10 m.g.d.).

The bulk of this installed capacity is likely to be distillation and currently MSF is the dominant process which has paved the way for desalination on the above scale to be even contemplated.

The larger-scale application of multi-stage flash techniques to sea-water distillation is due to two men: A. Frankel (formerly with Richardsons Westgarth & Co. Ltd., Northumberland) and quite independently but concurrently R. S. Silver (formerly with G. & J. Weir Ltd., Glasgow). (The desalination interests of both groups are now merged in a wholly owned Weir subsidiary, Weir-Westgarth Ltd., Troon.)

Credit is due to both of these men for realising the thermodynamic opportunities afforded by the flash distillation process, which had its roots in the old Alberger salt process for manufacturing salt from brine. Frankel [3] has stated the following in one of the most informative papers on the MSF process:

Flash evaporators have been known and built for nearly as long as submerged-coil evaporators and it would have been an obvious step to build multi-stage versions of them on similar constructional lines to those used for the submerged-coil evaporator. The following circumstances seem to have contributed to the fairly sudden emergence of the flash type:

1. In early practice, scale formation was accepted as inevitable in evaporators. In a submerged coil unit this resulted merely in a reduction of output, while the specific heat consumption remained substantially constant. In a flash evaporator, scale could cause not only a loss of output but also a deterioration in the specific heat consumption and an appreciable change in the flow resistance of the brine circulating system. Modern methods of sea-water treatment removed this disability from evaporators which can now be operated continuously over many thousands of hours, like chemical process plant. This improvement has been relatively more important to the flash evaporator.

2. A considerable increase in the unit size of evaporators has taken place very recently. Whereas five years ago 454 m³/day (100 000 gal/day) was about the largest unit size normally used, a considerable number of units of approximately 2 250 m³/day (500 000 gal/day) output are now in operation or under construction and there is no doubt that evaporator units of 4 546 m³/day (1 m.g.d) and more will be built before long. This trend has caused a considerable increase of the vapour volume to be handled inside a single unit. At the same time there occurred a considerable increase in the specific volume of the vapour, owing to the higher vacuum under which modern units have to operate. The larger the total volume of steam produced the more natural it is to let the steam condense on the outside of tubes, which are circulated by brine, rather than to resort to the opposite technique employed in the submerged-coil evaporator.

3. Designs have been evolved making possible the construction of flash evaporators with a large number of stages, but avoiding the need of using

separate shells for each stage, connected by costly inter-stage piping. The new designs have removed certain thermodynamics disadvantages which handicapped the flash principle and have produced a great improvement in the economics of evaporators.

In order to understand this leap forward more thoroughly the submerged coil evaporator is first analysed.

The submerged coil evaporator

A single-effect submerged coil evaporator as shown in Fig. 2.8 (p. 23) and consists of a shell, a steam-heating coil, a condenser cooled by cold sea water, part of which is bled off for the evaporator feed. Provision is made for distillate removal and brine blowdown to ensure that the evaporation brine concentration does not rise unduly.

In operation, the heating steam caused evaporation to take place from the brine surface. The vapour, which is pure water, condenses on the condenser tubes giving up its latent heat of evaporation. Part of the heated coolant is bled off and used as the evaporator feed and in this way some small energy economy is achieved as the feed requires less heating before evaporation commences. However, with this arrangement the steam consumption required to produce 1 000 kg water is roughly 2.33×10^9 Joules at 100°C (or 1 000 gal water requires 10^7 Btu at 212°F). Thus if boiler fuel were to cost £13.5 per ton, then the actual energy cost at the heating coil (neglecting boiler capital and maintenance costs) would be 50 p/10^9 J (50 p/10^6 Btu). Thus 1 m^3 distillate would cost £1.15 (1 000 gal would cost £5) with this single-effect arrangement based on energy consumption alone.

The incentives for multiple-effect operation can be clearly seen. If an evaporator could be designed with a performance ratio of 10 (i.e. 10 lb distillate per 1 000 Btu input) the energy cost per m^3 would be only £0.115 (1 000 gal would be only 50 p). Multiple-effect distillation was introduced with this aim in mind. If the single-effect evaporator of Fig. 2.8 is extended as shown in Fig. 4.1 to say three effects and the vapour from the first effect is led to the heating coil of the second effect and similarly that from the second to the third, then in theory, unit mass of vapour from the first effect will produce unit mass of vapour in the second effect which in turn produces unit mass of vapour in the third effect. Thus for the consumption of unit mass of steam from the boiler, a product rate three times that amount may be produced in a triple-effect evaporator – note that there is an implied relationship between number of effects and performance ratio in multiple-effect distillation, a point we shall return to later. In practice losses are such that 2.5 kg of distillate would be produced by the consumption of 1 kg saturated steam. For a multiple-effect system to operate, the pressure (and thus the corresponding saturation temperature) in the second effect is less than that of the first one so that the latent heat of vaporisation from the first effect may be effectively transferred through the heating coil of the second and succeeding effect. The practical restrictions of multiple-effect pool boiling systems, such as the one shown in Fig.

4.1, were such that six effects with a performance ratio of 4.9 was the maximum achieved – when VTE was developed to its current technical proficiency this restriction was removed as discussed in Chapter 5. The major design difference between multiple-effect evaporation and multi-stage flash distillation is that the

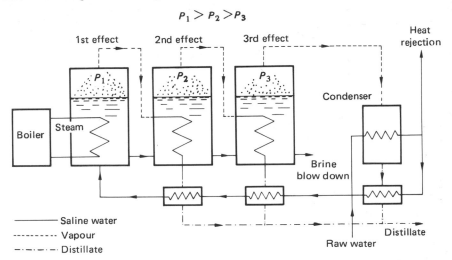

Fig. 4.1. Multiple-effect pool boiling distillation.

performance ratio is dependent on the number of effects in the former, but this does not necessarily apply in the latter, i.e. in MSF the number of stages ('effect' is not used in multi-stage flash terminology) is not a direct function of the performance ratio. Because of this it is perfectly possible, and in fact quite common, to obtain performance ratios of greater than 10 in multi-stage flash plants which was not possible with multiple-effect pool boiling systems.

Multi-stage flash principles

A single-stage flash plant as shown in Fig. 2.4 (p. 19) operates as follows. Sea water is heated in the brine heater to just below the saturation temperature t_{max} at P_{max}. It then enters the stage at reduced pressure P_1. The reduction in pressure causes the heated feed to 'flash' or commence evaporation to obtain thermal equilibrium with stage saturation conditions dictated by P_1. The reject brine is discharged and fresh sea water used as feed in this simple system.

The vapour formed in flashing condenses on the condenser or heat recovery tubes thus heating the incoming feed, to temperature t_{in}. The brine heater supplies the remainder of the energy required to raise the feed temperature to t_{max} before flashing commences. The fraction of brine which flashes in the first stage is given by

$$\frac{C(t_{max} - T_1 + \Delta_{BPE})}{L} \tag{4.1}$$

The assumption made in equation (4.1) above is that thermal equilibrium is attained in the stage – in practice this may not be so.

Now the single-stage flash plant in Fig. 2.4 (p. 19) can be extended to n stages as shown in simplified form in Fig. 4.2. The pressures in each stage are successively reduced until vapour volume, equilibrium and heat rejection considerations fix the minimum temperature of the last (nth) stage. For most purposes T_n is c. 38 43.5°C (100°F). Thus to increase the fraction of flashed vapour t_{max} must be increased. However, t_{max} is prescribed by $CaSO_4$ formation and is for most plants 122°C (250°F) and 93°C (200°F) for acid or polyphosphate-dosed plants, respectively.

As discussed in Chapter 2 the fraction of the brine stream which can be flashed is restricted to 0.1 to 0.15, i.e. to produce the desired rate of distillate requires a minimum brine circulation rate in the range 10–6.6 times the product rate. Thus a

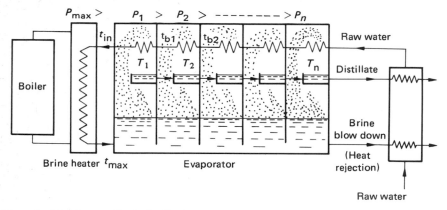

Fig. 4.2. Multi-stage flash distillation.

flash plant requires substantially larger brine flow rates than an equivalent multiple-effect plant. Not only that, other things being equal the mean temperature difference available for overall heat transfer in a flash stage is lower than the corresponding temperature difference in a submerged coil evaporator. In the flash evaporator the heat recovery raises the temperature of the brine in the condenser tubes which means a temperature rise in the circulating brine which in turns means a reduced temperature difference for heat transfer.

In the submerged coil evaporator the boiling brine temperature is a constant in each effect and so is the steam temperature; thus the temperature differences for heat transfer is constant. The superior thermodynamics of flash plants (*over those of multiple-effect pool boiling distillation*) could only be realised when the technology became available to allow the economic insertion of a large number of stage walls in the plant. A multiple-effect plant with 6 effects will have a performance ratio slightly less than 5, whereas a flash plant can have a performance ratio of 5 and use 12 or more stages. The thermodynamic characteristics of the multi-stage flash plant are illustrated by the following analysis based on a two-stage

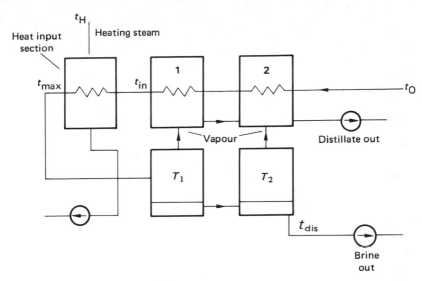

Fig. 4.3. Schematic two-stage flash evaporator.

flash plant, shown in Fig. 4.3, with the flash chambers schematically separated from the respective condensing sections.

For a given stage vapour temperature T_1 in flash chamber 1, the approach temperature of the brine leaving the condenser or heat recovery portion of stage 1 can be changed by varying the stage heat transfer surface. Increasing the surface reduces the terminal temperature difference $(T_1 - t_{in})$ by increasing t_{in} the brine temperature at inlet to the brine heater. Thus there will be an increase in the performance ratio proportional to

$$\frac{t_{max} - t_{dis}}{t_{max} - t_{in}} \qquad (4.2)$$

where $(t_{max} - t_{dis})$ scales the total product output, and $(t_{max} - t_{in})$ scales the thermal energy input to the brine heater.

A plot of temperature versus heat transfer surface areas can be drawn for the two-stage plant as shown in Fig. 4.4. The straight sloping line gives the profile for the feed in the brine heater and heat recovery sections. The stepped line gives the temperature profile for the flashing brine stream. The effect of varying heat transfer surface is shown. For a reduced heat transfer surface, temperature t_{in} is reduced and therefore there is a decrease in the performance ratio as scaled by equation (4.2). Note that even using an infinitely large heat transfer surface in each stage, the performance ratio cannot be increased above a certain maximum value which in the limit approaches the number of heat recovery stages. Thus for the two-stage flash plant in Fig. 4.3 the maximum theoretically attainable performance ratio is 2.

As long as the same design methods were considered and the same number of stages used as in multiple effect pool boiling, then the submerged coil units were superior. The flash evaporator rose to prominence when shell designs evolved that

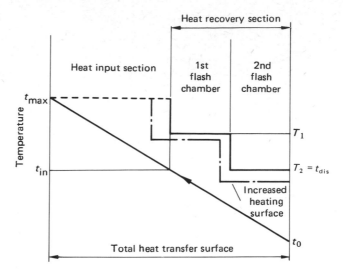

Fig. 4.4. Schematic temperature diagram for two-stage flash evaporator.

enabled much larger number of stages to be economically built than the submerged coil evaporator design. Thus an MSF plant with a performance ratio of 10 can be expected to have at least 20 stages (this is the basis of an MSF patent by Silver, [4] in that the number of stages is greater than twice the performance ratio). In practice a plant with a performance of 10 may have up to 30 to 35 heat recovery stages plus an additional 3 or 4 for heat rejection.

Stage number effect

Infinite number of stages

The effect of number of stages, is best analysed by considering the theoretical limit of an infinite number. Figure 4.5 shows a flow sheet with an infinite number of stages and brine recirculation employed as is customary for the evaporator feed. The corresponding temperature distribution for this arrangement is shown in Fig. 4.6. The heat rejection is a separate design entity and is obtained by circulating a separate cooling water supply through the last few stages, condenser tubes. The reject section tube surface area is dictated by the sea-water inlet temperature and flow rate as discussed in heat exchanger design in Chapter 2.

The feed M_f enters the bottom chamber and after blowdown M_d takes place at temperature t_{dis} to maintain a constant concentration on the plant the recirculated brine M_r is returned at stage $j + 1$ where the brine temperature t_{j+1} is apposite for heat transfer purposes. The important temperature profiles in Fig. 4.6 are the heavy black lines, the lower for the feed progression through the heat recovery stages and the upper for the flashing brine in both recovery and rejection stages.

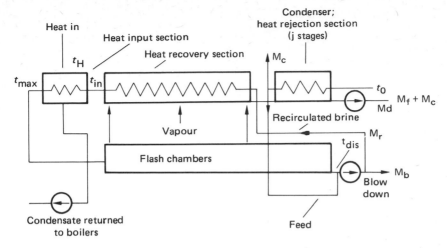

Fig. 4.5. Schematic diagram of a flash evaporator with an infinite number of stages.

Now, the brine and distillate are both allowed to flash in cascading down the stages and as the combined heat capacity of the flashing brine plus distillate streams equals that of the recirculating brine then the temperature rise of the recirculating brine equals the temperature drop of the flashing stream.

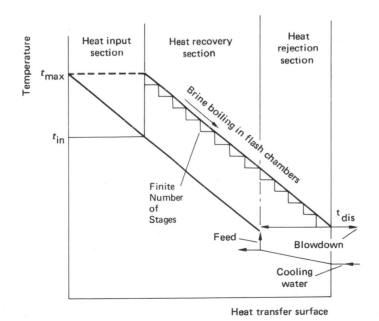

Fig. 4.6. Schematic temperature diagram for a flash evaporator with an infinite number of stages.

The performance ratio R is then given by

$$\frac{\text{lb prod}}{1\,000\text{ Btu}} = \frac{C(t_{max} - t_{dis})}{C(t_{max} - t_{in})}\,\frac{1\,000}{L} \qquad (4.3)$$
$$\text{steam}$$

as $L = 1\,000$ Btu/lb;

$$R = \frac{t_{max} - t_{dis}}{t_{max} - t_{in}} \qquad (4.4)$$

The brine circulation ratio r is given by

$$r = \frac{M_r}{M_d} = \frac{L}{C(t_{max} - t_{dis})} \qquad (4.5)$$

and the total distillate made may be calculated in kg (lb) with appropriate units used in (4.5) and (4.6)

$$M_d = M_r\,\frac{C(t_{max} - t_{dis})}{L} \qquad (4.6)$$

Now the designer can choose R independent of the number of stages. In doing so the heat transfer surface in the heat recovery sections changes thus:

$$A_s = \frac{L}{U(t_{max} - t_{in})}\ \text{m}^2\ \text{h/kg (ft}^2\ \text{h/lb)} \qquad (4.7)$$

(where $(t_{max} - t_{in})$ is the mean effective temperature difference for an infinite number of stages), i.e.

$$A_s = \frac{R \times L}{U(t_{max} - t_{dis})}\ \text{m}^2\ \text{h/kg (ft}^2\ \text{h/lb)} \qquad (4.8)$$

i.e. for a given set of conditions the recovery stage heat transfer surface area is directly proportional to the performance ratio – this is the same type of relationship as that for multiple effect evaporators with the significant difference that the number of stages can vary.

However, in all known plants the number of stages is strictly limited and the effect of a finite number is to reduce the mean effective temperature difference available for the maximum of $(t_{max} - t_{in})$ and thus a larger stage heat recovery surface is required to obtain the same performance ratio – the stepped line in Fig. 4.6 shows the effect of a finite number of stages.

Finite number of stages

The effect of a finite number of stages on the stage heat transfer surface is important and therefore a simplified derivation of the standard relationship for surface area in terms of performance ratio, number of stages and operating temperature range quoted in most technical papers is worthwhile:

We know $R = \dfrac{t_{max} - t_{dis}}{t_{max} - t_{in}}$

For a finite number of stages the temperature difference $(t_{max} - t_{in})$ depends on the minimum temperature difference Δt_m in each of the stages of the heat recovery sections and on the total number of stages (n), i.e. the temperature difference $(t_{max} - t_{in})$ has the following components: (neglecting boiling point elevation)

$$(t_{max} - t_{in}) = \Delta t_m + \frac{\text{flash range}}{n}$$

$$= \Delta t_m + \frac{t_{max} - t_{dis}}{n} \tag{4.9}$$

where it is assumed that there is an equal stage temperature drop.

The logarithmic mean temperature difference in each stage is now given by (from the relationships developed in Chapter 2) — where Δt_m is the minimum stage temperature difference and $(t_{max} - t_{in})$ the greatest.

$$\Delta T \log = \frac{(t_{max} - t_{in}) - \Delta t_m}{\log (t_{max} - t_{in})/\Delta t_m} \tag{4.10}$$

from (4.4) $(t_{max} - t_{in}) = \dfrac{t_{max} - t_{dis}}{R}$

from (4.9) $\Delta t_m = \dfrac{n(t_{max} - t_{in}) - (t_{max} - t_{dis})}{n}$

After rearrangement,

$$\Delta T \log = \frac{(t_{max} - t_{dis})}{n \log [n/(n-R)]} \tag{4.11}$$

Now the area required per unit mass of distillate made per unit time is (e.g. ft^2/lb/h or m^2/kg/h)

specific area $A_s = \dfrac{L}{U \Delta T \log} = \dfrac{L}{U} \dfrac{n}{(t_{max} - t_{dis})} \log \left(\dfrac{n}{n-R}\right) \tag{4.12}$

This neglects the effects of boiling-point elevation but does show the interrelationships of the design variables. Thus the effect of a finite number of stages may be readily obtained. One point to note is that to design a plant for a very high performance ratio can lead to high costs and an optimum has to be struck between capital and energy components.

Some of the parameters in equation (4.12) are fixed at the outset, i.e. U will be circa 2 830 W/m^2 °C (450–500 Btu/ft^2/h °F), and the flash range $(t_{max} - t_{dis})$ will be fixed by the method of scale control employed. Figure 4.7 shows a plot of equation (4.12) for $L = 2.33 \times 10^6$ 5/kg (1 000 Btu/lb), $U = 2\,830$ W/m^2 °C (500 Btu/h/ft^2 °F) and $(t_{max} - t_{dis}) = 55°$ and 83°C (100° and 150°F) respectively. The effect of increasing the flash range results in a substantial reduction in heat transfer surface.

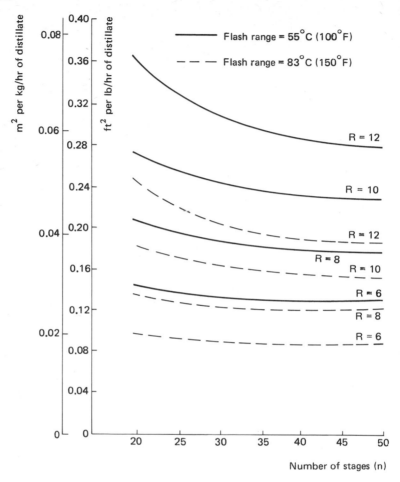

Fig. 4.7. Heat recovery surface area as functions of performance ratio, number of stages and flash range.

The designer then has to determine the evaporator performance ratio and the number of stages to be employed. Relationships such as equation (4.12) are useful in preliminary assessment of heat transfer surface in terms of the proposed number of stages. Other factors will enter that are not amenable to mathematical relationships, e.g. if the heat recovery tube lengths exceed a certain size a very sharp cost increase results; this may restrict n. As in all things the art of design is not a cut-and-dried process and allowances have to be made for losses which arise in practice.

Losses – necessary and otherwise

There are a variety of losses in an MSF plant that cannot be entirely eliminated. As discussed in Chapter 2 a drop in logarithmic mean temperature difference for heat exchange usually results.

The principal losses in this class are boiling point elevation and pressure drop loss. The sum total of these effects alone is estimated at $1.39°C$ ($2.5°F$) temperature reduction which has major implications for the heat transfer area required per stage. The effect is particularly marked in high performance ratio plants where logarithmic mean temperature differences are of order $3.9°C$ ($7°F$).

Other losses include poor venting resulting in vapour blanketing of the heat exchange surfaces which lowers the effective heat transfer coefficient and results in a reduced performance ratio. Tube fouling also falls into this class and can also significantly affect performance ratio. The significance of a lowered design performance ratio is not negligible, e.g. a $4\,546$ m^3/day (1 m.g.d.) plant designed for a performance ratio of 8 which in practice turns out to be 7 will incur an excess steam consumption of 0.816 kg (1.8×10^6 lb) steam per day. The cost of this excess steam consumption based on a fuel oil cost of £13.5 per ton is an extra £900 per day.

Equilibration

So far the condensation heat recovery process has been considered as the principal factor in plant design. However, distillation is a two-stage process involving both evaporation and condensation. The evaporation or flashing process characteristics can also have marked effects on plant performance ratio.

The governing characteristic of the flashing process is the achievement of complete equilibration. It is a subject which has received scant attention in the technical press as the design measures taken are kept within the individual manufacturer's ambience.

Basically, equilibration is the thermal equilibrium of the flashing brine stream with its surroundings in the flash chamber. Thus, if a stream of flashing liquid is at temperature t_{in} at entry to the flash chamber where the pressure is P_s, the flashing stream will eventually come to equilibrium at temperature t'_s where $t'_s = T_{sat} + \Delta_{BPE}$ corresponding to P_s. The attainment of equilibrium is not instantaneous and it is not unusual for the brine stream to leave the stage at an intermediate temperature t where $t'_s < t < t_{in}$. The equilibration achieved is characterised by a parameter termed the fraction of equilibration (β). where

$$\beta = \frac{t_{in} - t}{t_{in} - t'_s} \tag{4.13}$$

The flashing process is principally one of bubble growth within the mass of the liquid and has been analysed by Porteous and Muncaster. [5] The temperature of the vapour in the bubble (except in the early nucleation stage) may be taken as constant at the saturation value T_s. The driving temperature difference for evaporation across an element of surface may therefore be taken as $(t - T_s)$ whether the element represents part of a plane or dispersed interface. Thus we may write for the rate of change of t with time (θ):

$$C \frac{dt}{d\theta} = -hS(t - T_s)$$

or

$$\frac{dt}{(t - T_s)} = \frac{-hSd\theta}{c} \tag{4.14}$$

h is a suitably defined overall heat transfer coefficient assumed constant throughout the process.

Let $\qquad \bar{u}$ = mean bulk fluid velocity (constant)

$\qquad\qquad\quad L$ = flash chamber length

therefore $\qquad d\theta = \dfrac{dL}{\bar{u}}$ $\qquad\qquad\qquad\qquad\qquad\qquad\qquad\qquad$ (4.15)

Substituting (4.13) and (4.15) into (4.14) yields for $t_s \approx t_s'$

$$\frac{dB}{1 - \beta} = \frac{hS}{c\bar{u}} \times dL \tag{4.16}$$

Now assumptions have to be made about the bubble population which relates S to L. The net result is that (4.16) may be integrated to yield

$$\beta = 1 - e^{-n_e L} \tag{4.17}$$

Where n_e is a lumped parameter termed the evaporation index and is constant for any given set of flow conditions. The model predicts that flashing is an exponential decay process and therefore that the attainment of a high fraction of equilibration is very dependent on chamber length. The model has been tested on the data of Richardson Westgarth [6] for flash chambers and the results are shown in Fig. 4.8 as a semilog plot of $(1 - \beta)$ versus chamber length (feet) for the brine velocity employed in the tests.

The exponential decay character of flashing is immediately obvious as is the effect of stage temperature on equilibration rate attainment.

Also apparent from Fig. 4.8 is the appearance of an initiation length effect (L_0) before measurable flashing occurs. This is probably an entrance effect whereby the entrance weir to the flash chamber caused the submergence of bubble nuclei. (The Richardson Westgarth experimental run at 100°F was conducted with a different entrance geometry.) The flashing process may then be described by both an initiation length and a characteristic evaporation index. The flashing rate itself is determined solely by the evaporation index, the initiation length is a function of the duct geometry.

Table 4.1 below gives the measured value of n_e and L_0 for the data of Fig. 4.8.

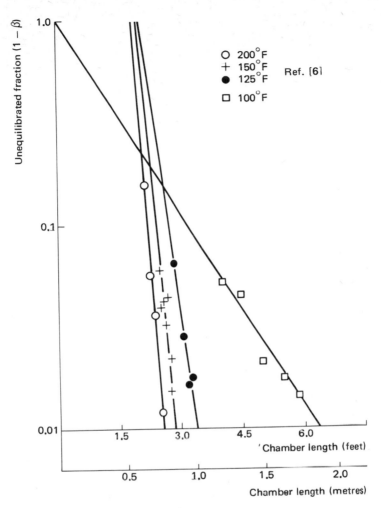

Fig. 4.8. Equilibration characteristics plot.

Table 4.1. Evaporation index for open channel flow from Richardson Westgarth data[6]

Brine temp. °C	°F	Evaporation = index (n_e) m^{-1}	ft^{-1}	Initiation length L_0 m	ft	Experimental run numbers Ref. [6]
37.8	100	2.31	0.706	0	0	19–24/6
51.5	125	12.0	3.66	0.63	2.07	15–18/6
65	150	17.4	5.34	0.607	1.99	14–54/4
93	200	20.4	6.26	0.562	1.845	1–7/6

It is seen from Table 4.1 that n_e is strongly influenced by fluid temperature and a principal reason for this is the hydrostatic head effect on the suppression of bubble growth. Table 4.2 below illustrates this aspect for the 38–71°C (100–160°F) temperature range where flashing problems are quite common. The data for a stage temperature drop of 2.78°C (5°F) which is the upper limit of most interstage temperature differences, for 1.39°C (2.5°F) the unsuppressed depth is approximately halved.

Table 4.2. Unsuppressed depth for 2.78°C (5°F) stage temperature drop

Brine temp.		Press difference for $\Delta T = 2.78°C (5°F)$		Head equivalent (cm/in of brine) unsuppressed depth	
°C	°F	Pa	psia	cm	in
38	100	1.05×10^3	0.152 4	10.8	4.25
49	120	1.72×10^3	0.249 6	17.5	6.9
60	140	2.7×10^3	0.392 4	24.2	9.5
71	160	4.09×10^3	0.594	42	16.5

The effect of hydrostatic head at around 38°C (100°F) is quite significant and as normal brine depths are spanned by the range 30–60 cm (12–24 in) considerable mixing must be carried out as bubble growth will only take place in the top 5–10 cm (2–4 in) of the flow. For this and heat rejection reasons final stage temperatures rarely fall below 38°C (100°F).

Equilibration has been shown to be a function of several factors chief among which is the circulation rate per unit width. A major constraint on this parameter is the occurrence of hydraulic jump which occurs at the critical velocity given by U_c $(\sqrt{gh})$. Thus for a brine depth of 30 cm (1 ft) the brine velocity must be less than 1.87 m/sec (5.7 ft/sec) otherwise flow instabilities result. The critical velocity leads to an upper bound on the brine flow rate per unit width which for 30 cm (1 ft) depth is 1.9×10^6 kg/h/m (1.3×10^6 lb/h/ft). Normally brine flow rates are restricted to a maximum of 1.46×10^6 kg/h/m width (1×10^6 lb/h/ft width). This flow rate can give rise to brine transfer problems at the low temperature end of the plant.

The design of flash chambers for correct equilibration can be a hit or miss affair. One quoted rule of thumb is for brine depth in the 30–40 cm (12–15 in) interval at 65°C (150°F) temperature and 1.66°C (3°F) flashing range is: 28–35 cm (11–14 in) of chamber length should be allocated per 1.46×10^3 kg/h/m (1 000 lb/h/ft) width of flow should be sufficient to attain equilibrium. Thus for a flow rate of 1.46×10^6 kg/h/m width (1×10^6 lb/h/ft) the chamber length could be 3.66 m (12 ft) and again to make sure at the low temperature end of the plant the user of this rule is told to make the stage length at least 6.1 m (20 ft). More will be said on flash chamber design later.

From the foregoing, it is seen that the flashing and condensation process are directly analogous. Each has an exponential decay nature – the former for the attainment of equilibration the latter for the attainment of temperature equality. It is in the nature of the condensation process inherently too expensive to operate in complete thermal equilibrium (i.e. regeneratively) and the designer makes allowances for this in his calculations. Similarly equilibration has inherent losses and both sets must be kept to a minimum to ensure maintenance of performance ratios.

Some recent flash-chamber dimensions have been such that values as low as 0.5 have been measured in the plant with a consequent reduction in performance ratio. When this happens plant instability occurs with surging and foaming leading to product contamination as the unequilibrated brine flashes at greater than the design rate in the downstream flash chambers. The exponential decay model developed above gives one possible approach to a better understanding of this problem.

Flash plant layout and construction

The basic layout of a flash evaporator plant is shown in Fig. 4.9. It consists of the following:

1. Feed treatment and degassing section
2. Feed heater section
3. Heat rejection section
4. Heat recovery section
5. Auxiliaries, pumps, ejectors, instrumentation.

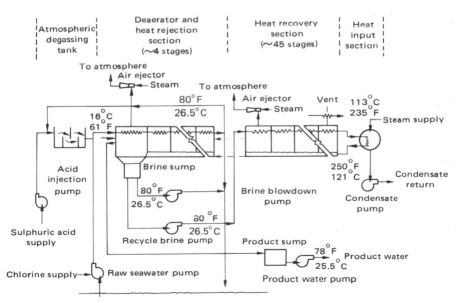

Fig. 4.9. Multi-stage flash distillation process flow sheet.

The major items of cost are heat transfer surface 34/40 per cent; shell and tube plates 30 per cent; pumps 15 per cent.

Heat recovery section

This is the major item of plant expenditure. We have already discussed the relationship of surface area, with the parameters under the designer's control, namely the number of stages and the performance ratio as given in equation (4.12). The effect of low number of stages, i.e. close to the value of R is to give a very high heat transfer surface. When n is much greater than R the heat transfer surface is greatly reduced but at the expense of a greater number of stages and hence interstage partitions. A compromise is reached between the competing demands which usually sets the number of recovery stages n_r approximately equal to $3R$.

The evaporator shell can be laid out in two very different ways – cross tube or long tube.

The cross-tube design, as its name implies, has the tubes transverse to the brine flow and incurs considerable fabrication work in terms of tube plates. For this design the heat exchange tube length fixes the stage width and hence the brine flow rate per unit width which in turn determines the stage length.

The long-tube design shown schematically in Fig. 4.10 has the tubes running parallel to the plant axis. The stage length is determined in this case by the length required for optimum heat exchange performance and the stage width is set to give the corresponding brine flow rate per unit width. The major advantages of this design are less tube plates required and considerably reduced pumping power due to the smaller number of water boxes with their attendant exit/entrance losses. It is claimed that maintenance is not so easily carried out in this design.

For plant capacities below 6 760 m^3/day (1.5 m.g.d.) the cross-tube and long-tube designs are equally favoured. Flash plants of UK manufacture tend to have cross tubes and those of US manufacture long tubes. For plants above 11 300 m^3 per day (2.5 m.g.d.) the long tube is considered to be superior as the tube length per stage and the chamber length are compatible and the tube bank height is much shallower than the cross tube which is desirable for good steam distribution and condensation rates.

Into the shell the designer must also fit appropriate orifices to control the brine flow rate and distribution. A splash plate is also provided to prevent contamination of the product. The vapour path needs consideration as the tube height above the brine surface determines to a large extent product purity and whether demisters are fitted or not. Normally wire-mesh demisters are inserted in the vapour flow path to disengage entrained droplets before condensation.

The condensate is collected in trays below the tube bundle and as it is at stage saturation temperature that it is allowed to cascade from stage to stage flashing in the process and is eventually discharged from the system at temperatures from 26.5°–37.8°C (80°–100°F). Because of the condensate flashing also (or more appropriately redistillation) the quantity of brine circulating to produce 1 kg

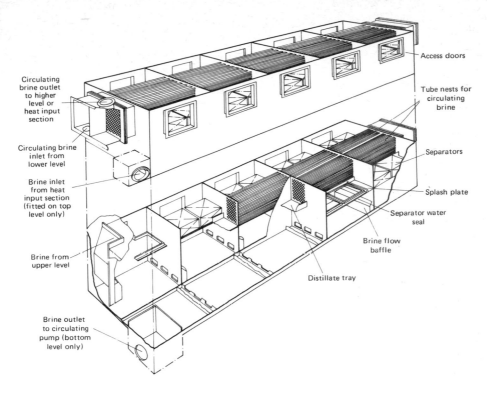

Circulating brine outlet to higher level or heat input section

Circulating brine inlet from lower level

Brine inlet from heat input section (fitted on top level only)

Brine from upper level

Brine outlet to circulating pump (bottom level only)

Access doors

Tube nests for circulating brine

Separators

Splash plate

Separator water seal

Brine flow baffle

Distillate tray

Fig. 4.10. Section of a horizontal evaporator – long tube design.

distillate is increased by 0.5 kg/kg distillate over the original value of L/C (flash range) i.e. the corrected brine circulation rate becomes:

$$M_r = M_d \left(\frac{L}{C(t_{max} - t_{dis})} + 0.5 \right) \text{kg/h} \qquad (4.18)$$

to allow for redistillation effects on output.

The stage height is determined by the sum of the following items:

1. Brine level + foam allowance
2. Disengagement height
3. Tube plate height.

Items 1 and 2 are similar for both cross- and long-tube arrangements as the constraints are common. Item 3 is of the same order of magnitude for both designs for plants in the 4 546–6 750 m³/day (1–1.5 m.g.d.) range.

However, above 11 300 m³/day (2.5 m g.d.) there can be significant savings in tube plate height with the long-tube design, especially if the tubes are distributed across the stage.

Heat rejection section

The function of the heat rejection section is as its name implies to reject the degraded thermal energy. As the evaporator does no work (other than the infinitesimal work of separation) all the higher grade energy input of the brine heater must eventually be rejected from the system. If the simplifying assumption is made of equal temperature drop in each stage of the plant then the ratio of recovery to rejection stages is the same as the performance ratio R which scales the condensation heat recovery to the heat input, i.e.

$$\frac{n_r}{n_j} = R$$

Thus for $R = 8$

$$n_r = 24$$

$$n_j = 3$$

Usually three or four rejection stages are used. Their design is along similar lines to the recovery ones except that vapour velocity considerations may dictate a much longer stage due to the high vapour specific volumes at saturation temperatures c. 43°C (100°F) and the need to keep the velocity low to avoid entrainment effects.

Disengaging rate consideration may also apply at the low temperature end. The maximum disengaging rate kg/h/m² (lb/h/ft²) of flashing surface for a given brine depth and flashing range varies with temperature. Thus at 43°C (100°F) the maximum disengaging rate is around 870 kg/h/m² (180 lb/h/ft²) and at 65.5°C (150°F) it increases to around 1 690 kg/h/m² (350 lb/h/ft²). The heat rejection section plant gives the designer many problems and eventually fixes the first stage temperature. The coolant arrangement for the rejection stages is indicated in Fig. 4.9. Raw sea water is chlorinated to c. 2 p.p.m. to avoid fouling and is circulated through the tube bank and if acid is used for scale prevention purposes, it is acid-dosed and sent to an atmospheric degasser for CO_2 stripping to 5–10 p.p.m. CO_2 content and then mixed with the recycle flashing brine stream in the first stage. The bulk of the dissolved air is then removed by the ejectors in this stage. Thus the heat rejection section performs the dual function of heat rejection and degassing. A fraction of the brine is drawn from the brine pump by the blowdown pump and is sent to discharge. The remainder is sent to the heat recovery stages by the recycle brine pump. The feed rate (M_f) is usually set at twice the product rate (M_d).

The mass balance for the flow streams is given by (4.19) as

$$M_f = M_b + M_d \tag{4.19}$$

The salt concentration is denoted by X plus corresponding subscript for the relevant stream.

The mass balance on salt

$$M_f X_f = M_b X_b + M_d X_d$$

where $X_d = 0$, as the product salt concentration is minimal

i.e.
$$\frac{X_b}{X_f} = \frac{M_f}{M_b} \tag{4.20}$$

As X_b must be limited by scale considerations and if X_f is normal sea water then for a 127°C (250°F) upper temperature limit,

$$\frac{X_b}{X_f} = 2$$

i.e.

$$M_b = \frac{M_f}{2} \tag{4.21}$$

Thus to maintain a concentration factor of 2 in the reject brine stream $M_f = 2M_d$.

The disposal of the heated concentrated effluent should be such that there are no recirculation effects near the plant intake which would effect the salt concentration or the turbidity of the raw feed. Other considerations which may rise to prominence in the future are the ecological implications of a heated effluent on esturial ecosystems. At least one challenge to the siting of a large US distillation plant has been on the grounds that copper ions in solution will adversely affect shell-fish – the projected large-scale adoption of desalination may require answers to questions in this area.

Pumps and auxiliaries

There are normally four main pumps in a distillation plant, namely recycle, raw feed, blowdown and product respectively. The principal cost item is the re-cycle pump which on a 4 546 m³/day (1 m.g.d.) plant can be required to circulate 1 800 tonnes of two-times concentrated sea water per hour. The type of pump chosen is a function of head and duties. However, as can be seen from the flow sheet this pump has both extraction and circulation duties in that it must extract the re-cycle stream plus deaerated make-up water from the brine stream and then circulate it round the heat recovery section. In order to perform the extraction duty properly without vaporisation occurring, a net positive suction head must be available and the pump must be sited below the evaporator in a well or on a lower level – which is a civil engineering function. An order of magnitude net positive suction head (NPSH) value is 10 m (33 ft) for this pump. The total head developed can be a *c.* 6.9×10^5 Pa (100 p.s.i.) which for a 1 800 tonne/h high-efficiency pump requires a shaft power of 410 kWh (550 horse power). A typical materials specification for this duty would be:

Casing	2Ni-resist
Impeller	type 316 stainless steel.
Shaft	type 316 stainless steel.

Continuous operation of 4 000 to 8 000 hours is required and the pump specifications must accordingly be rigorous.

The blowdown pump also requires a net positive suction head usually of order 4.2 m (14 ft). The materials specification for this pump are similar to the re-cycle pump, the head requirement is *c.* 1.375×10^5 Pa (20 p.s.i.) and for $M_b = M_d$ the capacity is 185 tonnes/h. The distillate pump has similar duties and net positive suction head requirements as per the blowdown pump. The materals specification is less rigorous. The feed pump has no NPSH requirement and can be a submerged type supplying coolant to both the rejection stages and air ejector condensers. Part of the reject coolant is used for feed and is sent for appropriate treatment as shown in Fig. 4.9 (p. 65).

The tendency towards larger plant designs may require the parallel operation of re-cycle pumps which may pose stability problems.

Air ejector

In order to extract the non-condensable gases released during the flashing process plus any air leakage each flash chamber must have provision for their removal. This is commonly done by cascading the non-condensables from stage and extracting at certain points in the plant. The ejector has an onerous duty, especially in acid-dosed plants, as CO_2 liberation brings about rapid corrosion and erosion of the venturi throats, diffusers and condenser tubes. For this reason, stainless steel is commonly employed for both diffusers and tubes. In operation the ejectors remove a mixture of non-condensable gases and vapour as dictated by Dalton's law of partial pressure. Any vapour loss must of course be made good at the brine heater. In order to keep the losses at a minimum the gases are extracted at the lowest possible temperature so that the ratio of non-condensables to vapour extracted is a maximum.

Brine heater

The brine heater is a conventional condenser with surface area requirements as discussed in Chapter 2. It is not normally an integral part of the shell. A typical materials specification is:

>   Tubes   90–10 cupro nickel;
>   Tube plates   carbon steel – clad or unclad

Water boxes are steel which can be lined with Cu-Ni plate. The overall heat transfer coefficients used in the brine heater design are higher than those used for the heat recovery surface evaluation as there are no non-condensables present and more efficient vapour distribution is also obtained. Values of 4 000–5 100 W/m² °C (700–900 Btu/ft² °F) have been reported by Foster and Herlihy [7] for the brine heater in the San Diego flash plant depending on the operating temperature and flow rates.

Plant operation

This aspect can require a book on its own, and it is intended here to only touch on a few of the problems experienced in plant operation. These problems to a great extent depend on whether the duty is continuous, as for example in Kuwait, or intermittent/standby, as for example in Jersey. The method of scale control employed is also important.

As discussed in Chapter 3 corrosion is a major hazard. This can be greatly influenced by the duty. Intermittent operation is conducive to air leakage which gives rise to substantial steelwork corrosion. It is not uncommon for up to 0.32 cm (0.125 in) iron oxide scale per year to form on unprotected mild steel and while this would not look particularly healthy it should give no trouble in operation unless detachment occurs. This can occur in intermittent duty plants where the scale can dry out, lose its adhesion and detach itself to arrive in the water boxes and foul the tube inlets. Water-box corrosion has been severe in standby plants due to this form of scale shedding, which is discussed by Clarke *et al.* [8] with reference to the Jersey plant. The corrosion aspects of continuously operated plants are much less serious and Stewart [9] has reported no known corrosion on a $3\ 400\ m^3/d$ (760 000 gal/day) installation which at the time had operated continuously for over one year. Similar experience has been reported by El Saie [10] on the continuous operation of flash installations in Kuwait.

Heat transfer surface fouling occurs in both acid- and polyphosphate-dosed plants. On the latter a soft easily removed sludge forms which can be removed at intervals determined by operational experience as discussed in Chapter 3. For acid-dosed plants alkaline and $CaSO_4$ scale formation can be prevented or removed by on-load acid cleaning at a pH around 2. However, a form of fouling termed copper oxide/iron oxide sludge deposition has been reported in acid-dosed plants [8] whereby a soft reddish-brown deposit forms at tube ends and the surrounding tube plate. This sludge has been removed by the circulation of soft rubber balls (Taprogge) through the tubes with a subsequent improvement in performance – Fig. 3.3 (p. 30) illustrates typical results from the Taprogge system.

Trouble-free intake systems are of major importance and many local plant difficulties stem from inadequate attention in this area. Material which has arrived in the inlet water can and has included marine vegetation, fish, sand and stones. Efficient screening and settling can keep most of these foreign bodies at bay. Frequently mussel or similar growths may occur in inlet pipes and chlorine shock dosing may be required to prevent fouling.

Foaming is a phenomenon which usually takes place in a few stages in any flash plant. Its occurrence is to be avoided as product contamination can result. The cause is often inadequate attainment of equilibration in the preceding flash chamber resulting in high vapour release rates in the next one due to the excessive brine temperature drop, e.g. for the Jersey plant (mean stage temperature drop $1.72°C$ ($3.1°F$)) the fraction of equilibration reported by Hawes [11] in some stages was as low as 0.5. The effect of this is to impose an initial driving temperature difference for flashing in the next stage downstream of $2.58°C$

(4.65°F) which could account for the observed violent fluctuations in the brine foam level. The attainment of controlled equilibration in each stage is fundamental for stable operation, and the maintenance of product purity. As with any departure from equilibrium there is a loss and the thermal energy degradation caused by excess flashing taking place at a lower temperature results in a lowered performance ratio.

Adequate venting of the dissolved air in the feed and the carbon dioxide released from bicarbonate decomposition plus any air leakage of those stages under vacuum

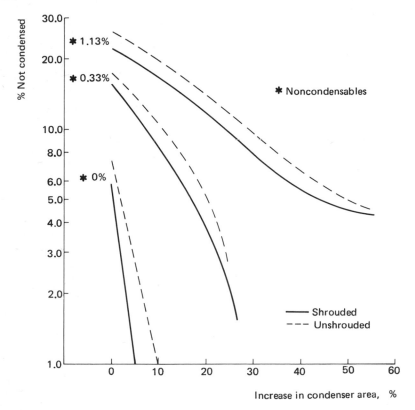

Fig. 4.11. Condenser performance plot as function of non-condensable content.

is essential to prevent blanketing of the heat transfer surfaces. Normal practice is to cascade the non-condensables from stage to stage and extract from the last stage. This practice is apparently satisfactory but calls for effective sweeping of non-condensables. The problem in multi-stage flash plant design is that a vapour velocity of *c*. 15 m/sec (50 ft/sec) is required for effective sweeping. For flash evaporators working up to 121°C (250°F) the high temperature stages operate at two atmospheres or more and the vapour flow velocities in the respective tube bundles are quite low. Non-condensable build up can occur and this aspect will require much greater attention in the larger plants currently contemplated. Figure 4.11 from

O.S.W. sources [12] shows the effectiveness of vapour removal as a plot of water vapour loss versus increase in bundle area. Where there are no non-condensables in the vapour stream, the heat transfer coefficient remains high and very little additional area is needed to remove the last of the water vapour. With 0.33 per cent air, condensing problems arise and the area required is substantially increased. The solid lines indicate the use of shrouds around the tube bundles to direct the vapour in a flow path best suited for concentration and removal of the non-condensables. For the data used in Fig. 4.11 the shrouded version has a definite advantage, at low stage pressures as in the heat rejection section the pressure drop created by the shrouds would militate against their adoption.

Summary

Multi-stage flash distillation is a reliable and proven process. It has market domination in terms of installed capacity and a strong following in the Gulf States where it will probably continue in use for many years. However, progress continues and the process superseded, namely multiple effect boiling has made a reappearance in the shape of long-tube vertical evaporation and experience to date suggests that significant product cost reductions are possible from the improved

Fig. 4.12. 6 700 m^3/day MSF plant, built by Weir Westgarth Ltd. for the Jersey Waterworks.

multiple effect process compared with an equivalent capacity and performance ratio MSF plant.

Fig. 4.13. 9 000 m³/day (2 m.g.d.) MSF unit built by Weir Westgarth Ltd. for Abu Dhabi – one of three identical units.

In order to visualise a completed MSF installation Fig. 4.12 shows the 6 700 m³/day (1.5 m.g.d.) plant supplied by Weir Westgarth Ltd. for the Jersey Waterworks. Figure 4.13 shows a Weir Westgarth plant which is one of three identical plants in Abu Dhabi.

References

1. Silver, R. S. 'Multi-stage flash distillation. The first ten years', Proc, 3rd Int. Symp. on Fresh Water from the Sea, Vol. 1, pp. 191–201, Djbrovnik, Sept. 1970.
2. Pugh, O. 'Desalination – its increasing role', Clean Water Conference, Adelaide, June 1972.
3. Frankel, A. 'Flash evaporators for the distillation of sea water', *Proc. I Mech. E.*, Vol. 174, No. 7, 1960.
4. Silver, R. S. British Patent No. 829,819.

5. Porteous, A. and Muncaster, R. 'An analysis of equilibration rates in flashing flow with particular reference to the multi-stage flash distillation process', Proc. 3rd Int. Symp. on Fresh Water from the Sea, Vol. 1, pp. 145–54, Djbrovnik, Sept. 1970.

6. Richardson Westgarth Ltd., 'Chamber geometry in multi-stage flash evaporators'. US Office of Saline Water Report No. 14-01-001-262, Nov. 1963.

7. Foster, A. C. and Herlihy, J. P. 'Operating experience at San Diego flash distillation plant'. Proc. 1st Int. Sump. on Water Desalination, Washington D.C., Oct. 1965, pp. 57–79.

8. Clarke, R. N. *et al.* 'Sea-water distillation plant, Jersey'. Some aspects of design and operating experience', Paper C16/73, I. Mech. E. Conf. on Water Desalination, London, Jan. 1973.

9. Stewart, J. M. 'Some practical aspects of desalination by evaporation', Proc. 1st Int. Symp. on Water Desalination, Washington D.C., Oct. 1965, pp. 651–67.

10. Ali el Said, M. H. *History, Experience and Economics of Water Production in desalination in Kuwait*, 1966, Vol. 1 pp. 77–195.

11. Hawes, R. I. 'Sea-water distillation studies in the U.K.A.E.A.', Paper C/18/73, pp. 33–41. I. Mech. E. Conf. on Water Distillation, London, Jan. 1973.

12. US Office of Saline Water 'Condenser tube bundle configurations for 50 m.g.d. and large desalination plants', O.S.W. abbreviated interim report prepared under the evaporator component development programme, 1970.

Chapter 5

Multiple-effect distillation

Introduction

The arrival of multi-stage flash distillation effectively eliminated the submerged tube multiple-effect process from the desalination field. However, multiple-effect evaporators are now in the ascendant being based principally on the long-tube vertical (LTV) falling film process which operates in an identical thermodynamic fashion to its predecessor.

The principle of the process, as illustrated in Fig. 5.1, is that the steam generated in the first effect is condensed in the second thus producing more steam which is cascaded to the third and so on. The principal advantage of the LTV process is that high heat transfer coefficients can be achieved with a considerable saving in heat transfer surface area. Also as the LTV process does not require the circulation of quantities of brine well in excess of the product output, shell costs are also reduced. Because of these cost advantages alone LTV development is now being vigorously pursued and it is now at the stage where it is a serious competitor to the multi-stage flash process. The savings claimed by Hawes [1] for a 2.27×10^4 m^3/day VTE

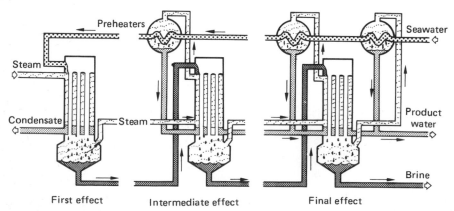

Fig. 5.1. Multiple-effect evaporation.

plant are 20 per cent reduction in product cost compared with the equivalent MSF plant operating under Mediterranean conditions.

VTE multiple-effect distillation — flow sheet

Figure 5.2 shows a typical flow sheet for a VTE plant. The arrangement utilises forward feed, i.e. the feed entering the highest temperature effect and flowing parallel to the vapour flow – this is the standard arrangement. The advantage of VTE *vis-à-vis* the MSF process are as follows:

1. Energy economy as the brine is not heated to above its boiling point as in the multi-stage flash process. This means that there are inherently less irreversibilities in the VTE process as the vapour is used at the temperature at which it is generated.
2. The feed is at its lowest concentration at the highest plant temperature hence scale formation risks are minimised. A higher maximum brine temperature or a smaller blowdown rate (or a combination of both) can be used.
3. The feed flows through the plant in series and as the maximum concentration only occurs in the last effect the worst boiling point elevation losses are confined to it.
4. The multi-stage flash process has a high electrical demand because of the recirculation pump.
5. Multi-stage flash is prone to equilibration problems which reflect themselves in a reduction in performance ratio. In multi-effect plants the vapour generated in one stage is used in the next and performance ratio is not subject to the whims of equilbration.
6. Plant simplicity is also promoted by the VTE process, e.g. a plant with a performance ratio of 10 may require 13 effects whereas a similar MSF plant could well use 30–36 stages. This entails a higher order of difficulties when brine, product and incondensables are to be transferred from stage to stage with minimum losses.

Referring to the flow sheet of Fig. 5.2 which shows the first three effects and the last effect only. The feed stream M_{FO} passes through the heat exchanger incorporated internally or externally for each effect. The feed flow is counter current to the vapour flow. The temperature rise gained by the feed in each effect is $\Delta T°C$ until it leaves the second effect heat exchanger and enters the feed preheater where it reaches the temperature t_{max} where t_{max} corresponds to the boiling point for pressure p_1 in effect number 1. The feed stream M_S is split into two components, M_{SOO} is used for preheating and M_{SO} is used for the first effect evaporator. The vapour formed in the first effect is then used as the feed steam in the second effect. The product from the first effect $M_{d1} = M_{D1}$ passes to a flash tank at pressure p_2 where a portion M'_{D1} flashes to attain equilibrium with the second effect pressure. The flashed portion is fed into the second effect – as the product flow rate increases with the number of effects the contribution of the flashing product to the thermal

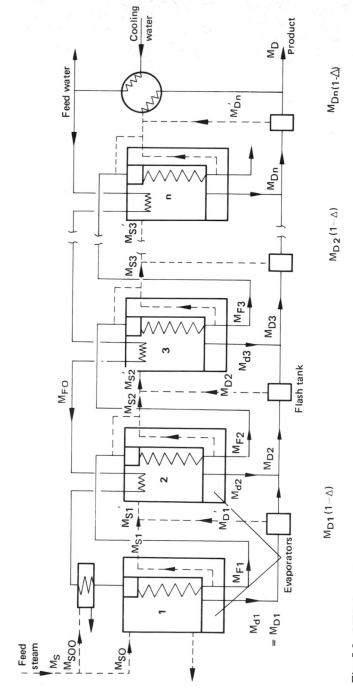

Fig. 5.2. VTE flow sheet.

economy of the plant is substantial. The blowdown from the first effect is the feed stream for the second effect, and so on, intereffect feed pumps are used for brine transfer. This process continues until the last effect is reached where the vapour is usually condensed in a separate heat exchanger.

Analysis of multiple-effect distillation

Consider the first effect. Mass and heat balances on the product and steam feed streams respectively shows:

$$M_{D1} = M_{d1} = M_{S0} = M_{S1} \tag{5.1}$$

and a mass balance on the feed stream shows:

$$M_{F1} = M_{F0} - M_{S0} \tag{5.2}$$

The feed stream to the second effect M'_{S1} is:

$$M'_{S1} = M_{S1} + M'_{D1} \tag{5.3}$$

where M'_{D1} is the contribution from the flashing of the product stream over ΔTx the temperature difference between the steam entering and leaving any effect

that is $\qquad \Delta Tx = \dfrac{\Delta T_T}{n} \quad$ and $\quad M'_{D1} = M_{D1} \times \dfrac{\Delta Tx}{L} \tag{5.4}$

let $\Delta Tx/L = k$, a constant for the plant.

Thus, $\qquad \begin{aligned} M'_{S1} &= M_{S1} + M_{D1} \times k \\ &= M_{S0} (1 + k) \end{aligned} \tag{5.5}$

but $\qquad M'_{S1} = M_{d2}$

and $\qquad M_{D2} = M_{d2} + M_{D1} (1 - k) \tag{5.6}$

Therefore $\qquad M_{D2} = M_{S0} (1 + k) + M_{S0} (1 + k)$

that is $\qquad M_{D2} = 2M_{S0} \tag{5.7}$

similarly $\qquad M_{Dn} = n M_{S0} \tag{5.8}$

that is the total product rate M_D is equal to the first effect feed steam rate times the number of effects. Thus the evaporation steam requirement is

$$M_{S0} = \frac{M_D}{n} \tag{5.9}$$

Now the total steam supply is

$$M_S = M_{S0} + M_{S00} \qquad (5.10)$$

where M_{S00} is the steam rate to the feed preheater whose duty is to heat the feed through $(\gamma_T + \alpha)°C$, i.e. through the terminal temperature difference of the effect feed heaters plus the temperature difference due to boiling point elevation.

The $\qquad M_{S00}L = M_{F0}(\gamma_T + \alpha)$

that is $\qquad M_{S00} = M_{F0}\left(\dfrac{\gamma_T + \alpha}{L}\right) \qquad (5.11)$

Now M_{F0} is usually set at $1.5\, M_D$ in an LTV plant, i.e. a blowdown concentration factor of 3 is the norm. Thus M_S can be rewritten as

$$M_S = \frac{M_D}{n} + 1.5\,\frac{(\gamma_T + \alpha)}{L}\,M_D \qquad (5.12)$$

and performance ratio $R = M_D/M_S$

that is $\qquad R = \dfrac{2nL}{2L + 3n(\gamma_T + \alpha)} \qquad (5.13)$

For a typical plant where

$$\alpha = 0.83°C\ (1.5°F)$$

$$n = 13$$

$$\gamma_T = 3.34°C\ (6°F)$$

$$L = 2.33 \times 10^6\ J/kg\ (1\ 000\ Btu/lb)$$

Then $\qquad R = 11.45$

Heat transfer area

Considering the first effect on its own as it has no feed heater, the condensing heat flux is $M_{S0}L$ and this is conveyed to the boiling brine through a temperature difference of $(\Delta T_x - \alpha)$

Thus $\qquad S_1 = \dfrac{M_{S0}L}{U_e(\Delta T_x - \alpha)} \qquad (5.14)$

where U_e is the mean evaporative overall heat transfer coefficient. Subsequent effects require the subtraction of the feed heating component flux $(M_{F0}\Delta T_x)$. Hence for each of the stages 2 to n the heat transfer area required is

$$S_{2-n} = (n-1)\left(\frac{M_{S0}L - M_{F0}\Delta T_x}{U_e\,(\Delta T_x - \alpha)}\right) \qquad (5.15)$$

and the total evaporative area is $(S_1 + S_{2-n})$. The feed heaters are assumed to have equal area and thus the total feed heater area becomes

$$\frac{n\,M_{FO}}{U_h} \log \left(\frac{\gamma_T + \Delta T_x}{\gamma_T} \right) \tag{5.16}$$

where U_h is the overall feed heater heat transfer coefficient. The total heat transfer area is thus given by the sum of equations (5.14), (5.15) and (5.16). Now for $M_{FO} = 3M_D/2$ and setting $\Delta Tx = \Delta T_T/n$ the specific area A_T/M_D becomes

$$\frac{A_T}{M_D} = \frac{n(L - \Delta T_T)}{U_e\,(\Delta T_T - n\alpha)}$$

$$+ \frac{1.5n}{U_h} \log \left[\left(\frac{\gamma_T + (\Delta T_T/n)}{\gamma_T} \right) \right] \tag{5.17}$$

Thus A_T/M_D can be obtained for fixed values of γ_T, ΔT_T, and n. L, U_e, U_h are fixed for any given plant though the selection of U_e is a design variable in that fluted tubes with a greatly improved U_e value over that of plain tubes may be specified.

From equation (5.17) it is seen that the specific area required is strongly influenced by the overall temperature range of the plant.

A comparison has been made by Burley [2] of the heat transfer areas required for both multiple-effect evaporation (LTV and fluted tube) and multi-stage flash processes respectively in terms of performance ratio, operating temperatures and heat transfer coefficients. The conditions used by Burley in his analysis are given in Table 5.1 (British units) and the results are shown in Fig. 5.3.

Figure 5.3 demonstrates why the LTV process has a superior chance of market penetration compared to MSF. MSF clearly requires more heating surface than LTV and the improvement with fluted tubes is even more significant. Obviously the use of a greater number of stages in MSF will reduce the area required, however the optimum number is roughly 3R. The development of fluted tubes has led to projections of 20 per cent reduction in product costs compared with MSF and it is with plants of this type that considerable development is now taking place.

Table 5.1. Comparison of MSF and LTV distillation processes

	MSF	*LTV*	*Fluted tubes*
$\Delta T_T(°F)$	150	150	150
$\alpha(°F)$	2.5	2.5	2.5
L Btu/lb	992	992	992
L_s Btu/lb	934	—	—
U Btu/h/ft² °F	500	—	—
U_e Btu/h/ft² °F	—	500	1 500
U_h Btu/h/ft² °F	—	500	500

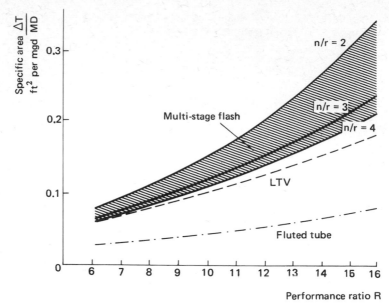

Fig. 5.3. Comparison of specific areas for MSF and LTV processes versus performance ratio.

Fluted tubes

The concept of enhanced heat transfer surface using thin film techniques is not new. Figure 5.4 shows a pictorial representation of a fluted tube VTE module. The incoming brine flows across an upper tube plate, through a distributor to form a film on the inside of the tube, which falls under the action of gravity. Steam from the previous effect or boiler condenses on the outside of the tubes forming the product distillate in a film. Heat transfer takes place across the condensate film, through the tube wall to the evaporating brine as shown in Fig. 5.5 which depicts the various resistances to heat transfer encountered. The two-phase mixture leaving the bottom of the tube is separated and the steam passed to the next effect.

The heat transfer resistance constraints are such that the tube wall resistance is fixed by the material used and its thickness which is dictated by design life. The designer then has to consider film resistances and modes of heat transfer which are, respectively, condensation on one side of the tube and boiling or liquid heating on the other. The principal resistance to heat transfer lies in the film thicknesses and many techniques have been tried to obtain thin film. The most successful, as it employs no moving parts or turbulence promoter, is the fluted tube which produces and maintains thin films in both the evaporation and condensation modes.

The effect of the flutes is to create surface tension forces inversely proportional to the flute radius of curvature which causes the condensate film to drain from the crests into the grooves. The result is that a substantial portion of the crest has a

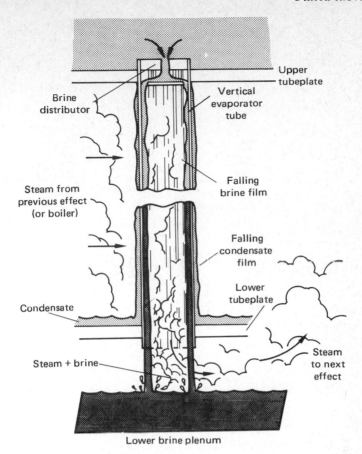

Upper
tubeplate

Brine
distributor

Vertical
evaporator
tube

Falling
brine film

Steam from
previous effect
(or boiler)

Falling
condensate
film

Lower
tubeplate

Condensate

Steam + brine

Steam
to next
effect

Lower brine plenum

Fig. 5.4. Fluted tube VTE module.

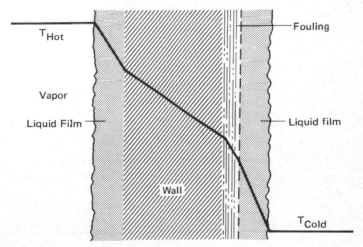

T_{Hot}

Fouling

Vapor

Liquid Film

Liquid film

Wall

T_{Cold}

Fig. 5.5. Heat transfer in the LTV evaporator.

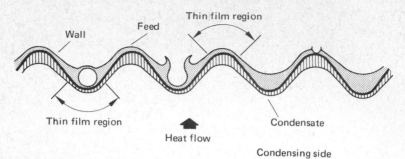

Fig. 5.6. VTE fluted tube heat transfer.

very thin film of condensate which enhances heat transfer in this area as illustrated in Fig. 5.6. The condensate in the grooves drains by gravity with reduced heat transfer in this area. The falling feed film also drains into the grooves under the action of surface tension. Boiling takes place in both crests and grooves and the crests are kept constantly wet by the boiling action in the grooves. The flutes also assist the feed distribution inside the tube and ensure uniform distribution along the length. The net result is that the high condensing and evaporating coefficients combine to produce overall heat transfer coefficients three to four times those of a

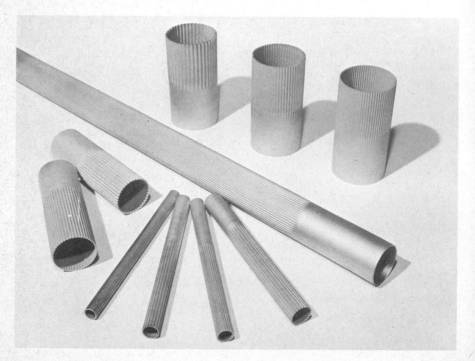

Fig. 5.7. Fluted tubes.

Fig. 5.8. Fluted tube heating chamber assembly being connected to vapour chamber sump on Gibraltar 1 360 m³/day plant.

plain tube. Tubes have been developed which can be readily manufactured and give coefficients of 1.4×10^4 W/m² °C (2 470 Btu/ft²/h °F) based on the original tube diameter. Approximate fluted tube costs are 30 per cent above the cost of a normal tube which is not significant compared to a three- or four-fold increase in overall heat transfer coefficient. Figure 5.7 shows one form of commercially available fluted tube and Fig. 5.8 shows a tube assembly being connected to a vapour chamber sump.

Flow distribution

In VTE plants, the film should be evenly distributed around the tube periphery and the brine should be equitably distributed to every tube in the effect. Maldistribution of the total flow could lead to the calcium sulphate solubility limit being exceeded in individual tubes causing scale deposition and a drop in performance ratio. Similarly uneven flow distribution leads to dry patch formation with a drop in performance ratio. The precipitation of calcium sulphate scale is to be avoided at all costs as its removal calls for severe treatment which in extremes means tube replacement.

Several varieties of flow distribution exist. One common method is to use a projection of the top edge of the tube as a weir and allow the brine to flow over. This has the disadvantage that fluctuations in head may cause maldistribution. Hawes [1] has reported the development of a nozzle shown in Fig. 5.9 which incorporates metering orifices that in turn form brine jets which hit a deflector and

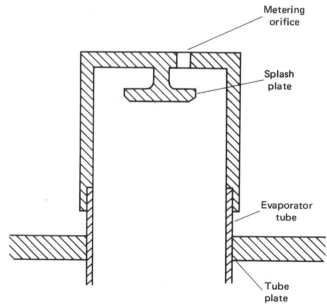

Fig. 5.9. VTE flow distribution nozzle.

thus produce a film on the tube inside walls. It is claimed that this nozzle arrangement is insensitive to head fluctuations and the brine flow was distributed to within ±10 per cent of the mean on a bank 68 tubes × 75 mm dia. × 95 mm triangular pitch.

Horizontal tube evaporator: multiple-effect evaporator (HTE)

One of the disadvantages of the VTE is that each effect requires its own pump to circulate the brine to the respective upper brine plenum. A possible multiple-effect arrangement which obviates this difficulty is the horizontal tube evaporator shown schematically in the two-effect arrangement in Fig. 5.10.

The principle of operation is exactly the same as in VTE, steam condenses on the inside of the tubes imparting its latent heat of condensation to the evaporating brine which cascades over the outside. The steam formed in the first effect is sent on to the next effect and the brine falls under gravity to feed the next effect, thus eliminating the intereffect pump. Effects can be stacked vertically and so only one pump instead of n is required for the brine.

The heat transfer rates in the HTE have been investigated by the Office of Saline Water and reported by Cox. [3] It was found that condensation inside the tubes was filmwise and that overall coefficients were approximately double those of smooth LTV tubes. Bubble nucleation on the evaporating side took place on the top of the tubes with rapid bubble growth and a sliding around the tube circumference taking place simultaneously. This mechanism caused rapid local rate of heat transfer to take place and it was concluded that the evaporative heat

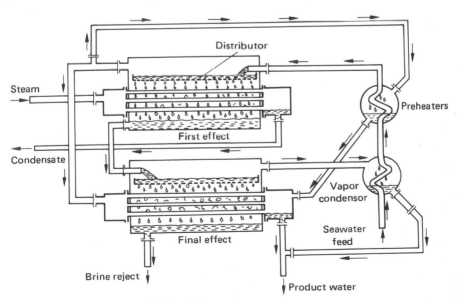

Fig. 5.10. Double-effect HTE module.

Fig. 5.11. Multiple-effect, horizontal, falling film plant showing feed distribution and boiling in progress.

transfer coefficient was independent of brine thickness but dependent upon the number and size of the bubbles. The principal resistance to heat flow was the condensing coefficient and in an analogous fashion to the use of fluted tubes in the LTV process it was found that a significant improvement could be obtained by the use of internal grooves which experimentation set at 0.254 mm x 0.8 mm (0.010 in deep x 0.031 in) spacing. The result was an overall HTE coefficient roughly three times that for an LTV plant using plain tubes which is a comparable improvement with that of fluted tubes in the latter process.

The operation of HTE was found to be very susceptible to the presence of non-condensable gases. Their blanketing effects were minimised by maintaining an exit velocity of greater than 2.45 m/sec (8 ft/sec) to sweep the gases out with the excess steam. The desired exit velocity is provided by locating the sea-water feed heater downstream of the effects, i.e. the steam condensed in any feed heater must first pass through the horizontal tubes of any effect.

The large-scale development of the HTE process is proceeding, Rhodes and Mills [4] have reported the successful operation of two test modules at Dungeness 'A' Nuclear Power Station. It is claimed that the HTE is the best distillation plant design when acid pretreatment is used as the formation of scale or tube attack and the boiling process can be readily observed through the inspection ports. Figure 5.11 shows feed distribution and boiling in progress in an HTE module.

Multiple-effect plant layout and operation

The layout of a multiple-effect plant is best illustrated with reference to Fig. 5.12 from the paper by Rhodes and Mills [4] (on which the following description is based) for the Gibraltar LTV fluted tube plant. This plant has 13 effects, a performance ratio of 10 and an output of 1 360 m³/day (300 000 gal/day). Equal area preheaters are used except preheater 13 which absorbs part of the last effect vapour. All heating chambers were made identical in size except for the last effect where a lower heat transfer coefficient required an increased tube length to provide the additional area. The heating chambers were mounted on common sumps, fabrication requirements dictating that four be used. The common sump technique enables brine and vapour to be separated. Brine feed and distillate streams can also flow from one effect to the next with the elimination of most of the inter-connecting pipework required if each effect were totally separate.

With the much smaller raw feed rates employed in ME plants, compared with MSF, great care has to be exercised in scale prevention measures if acid dosing is employed. The volume of sea water requiring treatment is substantially less, thus accidental overdosing is more probable which can lead to acid corrosion. The Gibraltar plant has automatic regulation by means of a feedback loop from a pH indicator on the degasser inlet which controls the injection pumps. Acid is diluted before use in order to suit the pump size. One interesting feature is the use of an acid blend box to ensure proper pH control and accurate pH readings at the sensor. The feed is extracted from the sea-water coolant stream for the excess vapour

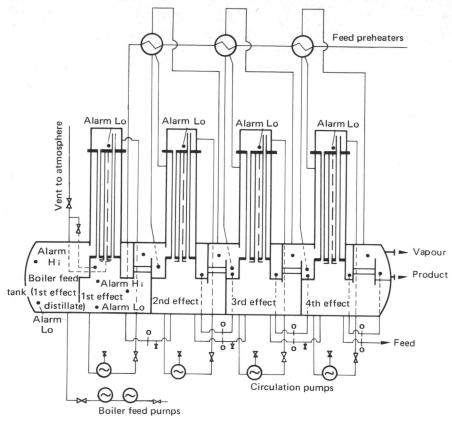

Fig. 5.12. Part of flow sheet for LTV sea-water distillation plant for the Government of Gibraltar.

condenser, it then passes in succession through No. 13 preheater, the ejector condenser and No. 12 preheater before acid dosing to pH 4.5. Degassing follows with stripping steam extracted from the 12th effect. After degassing the feed flows through 11 U-type preheaters in series. After the feed has passed the preheater chain it enters the No. 1 effect sump where it is recirculated to the top of the heating chamber.

Each heating chamber has a top water box fastened to the tube plate which serves as the brine plenum. Recirculated brine passes through a wire mesh to prevent scale clogging the tube distributor orifices and flows down the inside of the fluted tubes. Non-condensable gases are extracted through a central tube with orifices along its length. The condensate from the first effect cascades into an extension of the sump which acts as a boiler feed tank. The vapour separates from the brine which is then transferred to the second effect sump through an orifice and recirculated. Vapour from the preheater is extracted from the top of the heating chamber so that velocities are maintained throughout its length.

The steam is supplied by a packaged boiler and is used in a 200 kW turbo alternator before exhausting to the No. 1 effect. Part of the boiler steam is used for the ejector and a dump condenser is available for balancing the system during start up and shut down.

The materials of construction are similar to those for MSF plants. Main effect sumps are carbon steel with a large corrosion allowance. All water boxes are aluminium bronze except those on the heating chambers where low velocities are employed and carbon steel is therefore satisfactory. Tubes are aluminium brass except for the ejector condenser and non-condensable gas cooler where titanium is employed because of the corrosive duties. All pipework is carbon steel except sea-water services before the degasser which are PVC or rubber-lined steel. The degasser is rubber-lined carbon steel.

Scaling in VTE plants

As with any other distillation process, scale prevention is of paramount importance. Experimental work by Hodgson *et al.* [5] shows that the $CaSO_4$ prevention guidelines discussed in Chapter 3 apply, with one proviso. Heat flux as well as temperature and concentration must be taken into account in the VTE falling film evaporator. The results of Hodgson *et al.* show the importance of heat flux on $CaSO_4$ formation, e.g. at 'low' heat flux $CaSO_4$ light deposition was encountered at $130°C\,(266°F)$, whereas a 'high' heat flux deposition was observed at $115°C\,(240°F)$. With high heat flux typical experimental values were

Concentration factor   1.05
Temperature $115°C\,(240°F)$
Heat flux $7 \times 10^{-4}\ W/m^2\ (23\ 000\ Btu/h\ ft^2)$
CaSO₄ formation recorded

with low heat flux

Concentration factor   1.0
Temperature $130°C\,(266°F)$
Heat flux $3 \times 10^{-4}\ W/m^2\ (10\ 000\ Btu/h\ ft^2)$
Light CaSO₄ deposition

The work further showed that at a given temperature, scaling of the heating tube surface occurs at a lower concentration in VTE compared with MSF. This is explained by the presence (presumed) of areas of high concentration forming within the film as a result of uneven evaporation. This work enables VTE plants to be designed for high temperature operation.

Present position

Both VTE and HTE systems are commercially available. VTE fluted tube ship-board installations have been in use for several years. A commercial VTE fluted tube

Fig. 5.13. 1 360 m^3/day (300 000 gal/day) VTE plant for Gibraltar.

Fig. 5.14. 4 282   m^3/day (942 000 gal/day) VTE installation at Europort Rotterdam.

installation with a performance ratio of 10 and 1.36×10^3 m^3/day is in operation in Gibraltar with a claimed reduction in water costs compared with an equivalent MSF plant. The improved multiple-effect processes as embodied in the LTV and HTE concepts will assume growing importance in water supplies by desalination in the future. Figures 5.13 and 5.14 show two VTE installations completed by Aitons Ltd. for Gibraltar and Europort Rotterdam, respectively.

References

1. Hawes, R. I. 'Sea-water distillation studies in the U.K.A.E.A.', Paper C/18/73, pp. 33–41. I. Mech. E. Conf. on Water Distillation, London, Jan. 1973.
2. Burley, M. J. 'Analytical comparison of the multi-stage flash and long tube vertical distillation processes', 2nd European Symp. on Fresh Water from the Sea, Athens, May 1967.
3. Cox, R. B. 'Some factors affecting heat transfer coefficients in the horizontal tube multiple effect (HTME) distillation process', Proc. 3rd Int. Symp. on Fresh Water from the Sea, Vol. 1, pp. 247–63, Djbrovnik, 1970.
4. Rhodes, C. and Mills, K. E. 'The Gibraltar multiple-effect VTE falling film plant', Paper 6/19/73. I. Mech. E. Conf. on Water Distillation, London, 1973, pp. 43–49.
5. Hodgson *et al.* 'Calcium sulphate scaling in falling film evaporators', Proc. 4th Int. Symp. on Fresh Water from the Sea, Vol. 2, pp. 143–59 Heidelberg, 1973.

Chapter 6

Miscellaneous processes

Introduction

This chapter embraces a few minor processes each with their own sphere of applicability. The first is vapour compression distillation where the energy input is supplied by a compressor instead of a heat exchanger. Next is solar distillation which relies on solar energy for its operation thus entailing different design concepts from conventional methods. The opportunity will also be taken to discuss within the umbrella of this chapter one of the fringe developments in the distillation field, namely low temperature difference distillation which may grow in importance as energy costs increase.

Vapour compression distillation

Figure 6.1 shows schematically the principle of vapour compression distillation. The inlet feed water is heated initially in liquid/liquid heat exchangers by the blowdown and product streams respectively then passes to a brine heater where it is heated by vapour which has been discharged by the evaporator compressor at a temperature (T_{so}) greater than T_s the evaporator vapour temperature. The heated feed passes into the evaporator where flashing takes place. The vapour released is compressed and used as the steam supply for the brine heater and the condensate discharged as the product stream. Figure 6.1 shows the compressor coupled to a flash chamber, it can also be linked to a distillation effect. Multiple-effect operation is possible as the steam generated in the last effect may be compressed and used as the feed steam to the first effect as shown in Fig. 6.2.

The advantage of the vapour compression process is that the latent heat of the vapour released by evaporation is in a sense 're-cycled'. The compressor supplies the energy input necessary for the flash drop plus losses, i.e. it may have to compress the vapour over a 55°–67°C (10/12°F) temperature rise. Thus the energy input per unit mass of product is low and very high performance ratios (in excess of 15.1) can be obtained. The only large-scale vapour compression plant is an 3.750 m³/day

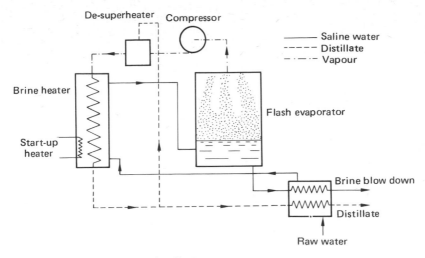

Fig. 6.1. Vapour compression distillation.

(0.83 m.g.d.) installation at Rosewell, New Mexico. The energy requirement of the Rosewell installation is around 16.6 kWh/m³ (75 kWh/1 000 gal) but it must be stressed that this is a one-off installation and the use of improved heat exchange methods have projected *c.* 11.1 kWh/m³ (50 kWh/1 000 gal) as being attainable by the vapour compression process. Silver [1] has drawn attention to the need to compare energy requirements on a common basis, as the production of 1 kWh electrical means the consumption of roughly 3 kWh thermal. The only proper basis for comparison is what is the energy cost per m³ (or 1 000 gal) product? In the example used in Chapter 4 the energy cost for multi-stage plant with a performance ratio of 10 using fuel oil at £13.5 per ton was 11 p/m³ (50 p/1 000 gal). For

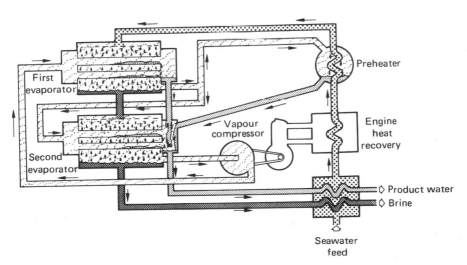

Fig. 6.2. Multiple-effect vapour compression.

electricity at an off-peak price of 0.6 p/kWh the energy cost for vapour compression at a consumption of 16.6 kWh/m³ (75 kWh/1 000 gal) is 10 p/m³ (45 p/1 000 gal) which is a potential saving on this component of the water cost but these conjectural savings are obviously totally dependent on the power cost. However, the vapour compression process has never been developed on a large scale

Fig. 6.3. Packaged 4 000 g.p.d. vapour compression plant.

with the exception of the Rosewell plant. It is capital intensive due to the high cost of compressors and liquid/liquid heat exchangers which has tended to keep its application confined to the 4.5–90 m³/day (1 000–20 000 gal/day) range.

Various proposals have been made to improve the efficiency of the VCE process and reduce its high capital costs. These centre round the use of vapour compressor multi-stage flash systems as proposed by Starmer and Lowes [2] and Wood and

Herbert [3]. The combination of vapour compressors and multi-stage flash allows the use of a compressor which handles as vapour a small proportion of the product output and the use of higher compression ratios. This enables a true compressor to be used as distinct from the blower designs commonly employed for the small compression ratios in single-stage or single-effect plants. Efficiencies are increased as losses are minimised due to the reduced vapour flow rates compared with a single-stage plant.

To date there is no indication that these proposals will be taken up and at the moment the main market for vapour compression plants lies in the 4.5–90 m^3/day (1 000–20 000 gal/day) class as illustrated in Fig. 6.3.

Vapour compression plants are often specified for those duties not covered by the range of larger installations which means that they may service, say, a Western Australia whaling station or some remote island where supplies in the above range are required. The maintenance accorded may be indifferent and feed-water treatment may be neglected and perhaps for these reasons the process has not had high marks for reliability so far, with the notable exception of the US Rosewell installation which has performed successfully for several years. An Australian Government report [4] gives a detailed breakdown of the problems encountered with vapour compression installations and itemises among others that one plant required its heat exchanger tubes to be replaced after three months due to faulty scale-control techniques. Another plant required tube replacement after six months due to scale formation in the evaporator tubes. The selection of this process requires careful consideration of the raw feed composition and the quality of labour available for maintenance. It is often employed inland on borehole waters with high concentration of Ca, Mg, SO_2 and HCO_3. In such cases the process choice may be very restricted and solar distillation may be specified instead as was done at the Australian opal mining town of Coober Pedy where the raw feed is virtually saturated at borehole conditions with scale-forming constituents.

Solar distillation

Solar distillation utilises, in common with all distillation processes, the evaporation and condensation modes but the resemblance ends there. It is a technique with no capital resource energy requirements needing only an adequate supply of solar energy. It can tackle raw feed of any composition and can be a viable proposition for outputs up to roughly 45 m^3/day (10 000 gal/day) or greater depending on location. Its use is mainly confined to areas with high solar radiation intensities (which embraces all the world's arid zones), where fuel is expensive and skilled labour scarce or non-existent. The extreme simplicity and reliability renders it suitable for use in many developing countries.

Principle of solar distillation
Solar distillation is based on the 'greenhouse effect' whereby glass and other transparent materials have the property of transmitting incident short-wave solar

radiation. Any surface underneath the glass cover can then be heated. However, the re-radiated wavelengths from the heated surface are such that little radiant energy can be transmitted back through the glass and hence there is a temperature rise in the enclosure. This effect is put to use in solar stills to produce fresh water from brackish or saline water by evaporation, energy being supplied by the incoming solar radiation.

The mechanism of solar radiation transmission through glass is shown schematically in Fig. 6.4. It is seen that the bulk of the solar radiation is transmitted directly through the glass. The net solar heat gain as a percentage of incident radiation is given in Fig. 6.5, based on data from Pilkingtons. [5] For

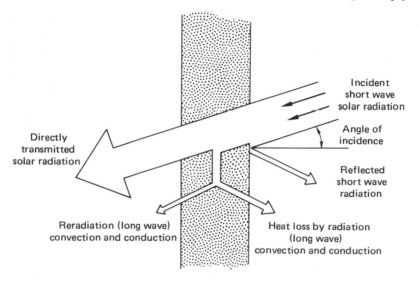

Fig. 6.4. Mechanism of solar radiation transmission through glass.

angles of incidence up to 35° the total solar heat gain is as high as 85 per cent of incident radiation. The areas of applicability of solar distillation are roughly banded by latitudes 35°N to 35°S which embraces all the world's arid zones. There is obviously a wide range of average solar radiation intensities in this band and Table 6.1 from Lof [6] gives the monthly and average values (cal/m^2/day) for various locations throughout the world. Conversion to W/m^2 = insolation (4.83 x 10^{-5}).

Still output per unit area is determined by solar radiation intensity – all other things being equal. Thus from Table 6.1 a solar still located in Alice Springs would have a much greater output than one located in Cambridge, UK – although solar water heaters, based on the greenhouse principle, for swimming pools are sold in the UK.

Still construction

Still construction is or should be a simple affair, as sophistication is not desired or desirable in the usual geographical locations where this method of desalination is

Table 6.1. Selected values of solar radiation at 10^4 cal/m²/day

Location	Lat	Long	elev (m)	Jan	Feb	Mar	Apr	May	Jun	Jul	Aug	Sep	Oct	Nov	Dec	Ann'l
Aden	12°50'N	45°01'E	4	415	481	541	591	573	530	503	535	541	544	500	444	516
Australia																
Alice Springs	23°48'S	133°53'E	546	646	620	556	466	369	337	361	460	551	592	623	641	518
Dry Creek, SA	34°50'S	138°35'E	4	691	601	487	426	233	197	209	286	397	479	600	662	439
Ceylon																
Batticaloa	07°43'N	81°42'E	3	440	510	550	540	540	510	520	530	540	500	470	420	507
France																
Montpellier	43°35'N	3°50'E		150	230	300	420	480	600	640	490	380	240	170	120	352
Greece																
Athens	37°58'N	23°43'E	107	180	276	334	457	516	577	572	498	396	276	187	161	369
India																
Baroda	22°15'N	73°15'E		450	520	580	650	690	600	450	400	540	470	480	420	521
Kenya																
Nairobi	01°18'S	36°45'E	1 799	558	595	559	486	416	397	324	366	464	492	486	522	472
Peru																
Huançayo	12°02'S	75°19'W	3 313	670	518	574	540	499	491	519	568	597	634	629	596	570
Spain																
Almeria	37°N	2·5°W		215	296	403	503	552	588	595	541	443	337	241	190	409
UK																
Cambridge	52°13'N	00°06'E	23	60	102	190	284	398	428	412	320	246	151	67	42	225
USA																
Boston	42°21'N	71°04'W	102	139	198	293	364	472	499	496	425	341	238	145	119	311
Phoenix	33°26'N	112°01'W	102	297	408	521	643	724	740	652	612	568	452	339	280	520

Conversion to W/m² = insolation (4.83×10^{-5})

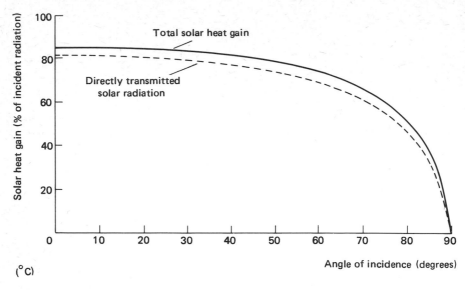

Fig. 6.5. Net solar heat gain as a percentage of incident radiation.

used. A land-based solar still installation essentially comprises a brine pool with associated radiation-absorbent liner, a supporting structure and a glass or plastic cover. Provision is made for the introduction of feed water, removal of distillate and disposal of reject brine. Storage of the product may also be required. The brine is usually contained in a level quiescent pool, hence the name 'basin-type' still is given to this form of installation. Continuous flow may be obtained by providing a gentle gradient on the base and allowing the feed to trickle in and the reject brine out.

Figure 6.6 shows a cross-section of a glass-roof basin still constructed by

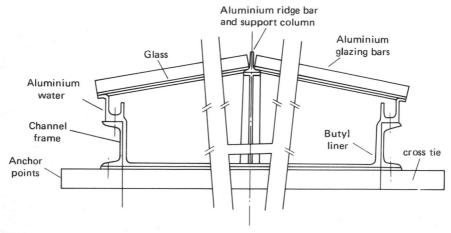

Fig. 6.6. Basin-type still.

Porteous [7] for the Royal Society Research Station on the island of Aldabra in the Indian Ocean. This was commissioned in July 1970. The design is based on a series of standard prefabricated modules which were readily erected on the island by local fishermen. Each module has a brine pool area of 42 m². The Aldabra installation has four such modules to provide a reliable means of water supply on this remote coral atoll. Rainwater catchment gutters are also provided, thus augmenting the still output considerably in wet weather when solar radiation may be insufficient for significant distillate output.

In operation, the incoming solar radiation heats the brine pool and evaporation takes place when the brine reaches the temperature range 50°-65°C (122°-149°F). As the roof is transparent to solar radiation it is below the saturation temperature of the mixture of air and water vapour in the enclosure and condensation takes place on the sloping surface. The condensate forms in drops or rivulets, and because of the wetting characteristics of the glass it runs into condensate channels, whence it is led out of the system. The normal mode of operation is to allow evaporation of the brine to a concentration of twice that of the incoming feed and then drain and refill. In most stills this is done once every three to seven days, depending on depth of brine pool and insolation.

Several points are worth mentioning. Many solar still designs have been mooted Bloemer *et al.*; [8] Howe and Tleimat; [9] Morse [10] most with emphasis on cheap construction through the use of plastics for the roof and very light supporting structures, often with polythene-lined brine pools. In the writer's opinion, which has been confirmed in practice, there is no substitute for glass, a well-supported structure, and a heavy gauge pool liner if the object of the design is long life, reliability and ease of maintenance. These statements are borne out by the fact that the Symi still constructed by Delyannis and Piperoglou [11] has had its plastic roof replaced and their subsequent Patmos installation used glass instead of plastic. In Australia, prototype continuous stills have met with difficulty in performance maintenance through ruptured plastic liners and other cooling problems. There are two inherent problems with plastics. They possess adverse wetting characteristics, thus condensate drainage is poor and the subsequent 'fogging' impedes the transmission of solar radiation. Their life is also short under the high radiation conditions normally prevalent as decomposition takes place under the action of ultraviolet radiation.

A form of continuous still has been under development in Australia for about ten years, but little information is available on the current design or its performance, but the Mark II (1966) design has been documented. [12] The major problem with continuous distillation is the control of a very thin film of saline water as it flows down a slight incline (1 in 60) under a glass covered enclosure some 20 m long. If 'dry patching' occurs on the pool liner, a hot spot develops and scale build-up is rapid.

Basin-type stills are relatively simple affairs as shown in Fig. 6.6, and it is not surprising that they form the majority of the world-installed solar still capacity.

The output of a still is determined by the intensity of solar radiation and the area covered, all other things being equal. Table 6.2 gives details of a few

Table 6.2.

Location	Brine pool area (m^2)	Date of erection	Type	Average annual distillate output (based on pool area) $l/m^2/day$	Corresponding (average insolation) $kcal/m^2/day$	W/m^2	Data reference
Indian Ocean (Aldabra Is.)	167	1970	Basin	3.8	5 160	250	Author's data from Aldabra Research Station
Chile: Andes (Elevation 1 300 m)	4 700	1872	Basin	5.76 (max.)	8 150	395	Harding (1883)
Greece:							
Patmos	8 667	1967	Basin	3.0	Not given	–	Delyannis and Piperoglou (1967)
Symi	2 700	1964	Basin	2.62	3 743	182	Delyannis and Piperoglou (1968)
Australia:							
Coober Pedy	3 500	1966	Continuous	3.22	5 100	246	Morse (1967)
Muresk	416	1936	Continuous	2.2	5 100 (author's est.)	246	Morse (1967)
Pacific Is.	4·65	1966	Basin (plastic roof)	4.3 (max)	6 300 (author's est. from Lof 1966)	305	Howe and Tleimat (1967)

installations where published figures are available. It is noted in passing that there is in principle little difference in construction between a 50 m^2 and a 25 000 m^2 still. The relevant parameters by which performance can be judged are average insolation and output per unit area and these are given in the table. An order of magnitude for a well-designed basin still is 51 m^2/day (1 gal/day 10 ft^2) for an insolation of 340 W/m^2 (7 000 kcal/m^2/day).

Performance

The theoretical performance of solar distillation has been analysed by Porteous [13] and is mainly a function of insolation and base loss, i.e. the heat losses through the base have a strong effect on still output. The results of the theoretical

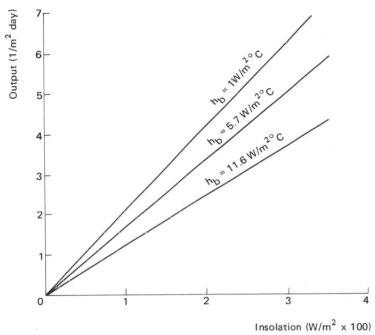

Fig. 6.7. Variation of output with insolation.

analysis are summarised in Fig. 6.7 which shows still output plotted against insolation with base loss (h_b) as parameter. It is seen that the effect of h_b is quite marked at high insolation, e.g. at 300 W/m a reduction in h_b from 5.7 to 1 W/m^2 °C will give an approximately 20 per cent increase in output. A typical loss coefficient for a base of sand and gravel is 5.7 W/m^2 °C.

Practical comparison is possible as the performance of several solar still installations have been monitored; Symi (Delyannis and Piperoglou [10]), Coober Pedy (Morse and Read [12]) and Florida (Bloemer *et al.* [8]) respectively. Data on the Aldabra installation was also collected by the research station staff and this along with the data for the Symi, Florida and Coober Pedy stills is shown in Fig.

6.8 as a plot of output versus insolation. It is seen that the performance of the Aldabra, Florida and Symi stills are quite similar. This is to be expected as these were basin-type solar stills. The Coober Pedy installation was continuous, and again, as would be expected, the performance is substantially less due to the characteristics of these plants.

Over the usual operational range of 200 to 350 W/m^2 there is a good measure of agreement between the predicted and actual outputs. Confidence can therefore be taken in the theoretical predictions given that the base loss coefficient can be estimated for the installation. The main utility of the theory, however, is to demonstrate the influence on output of the various parameters under the designer's control.

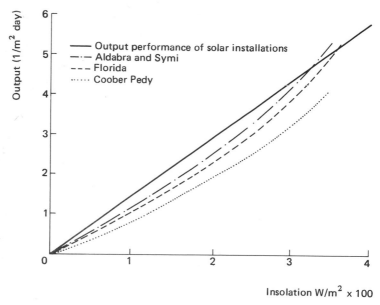

Fig. 6.8.

Solar distillation plants are finding favour with many isolated communities and can successfully deal with borehole water on which many distillation plants are unable to operate. The process is not usually a solution where skilled labour and cheap fuel are available but it is capable of great dependability for the provision of water at a cost comparable with other processes in its range of applicability. Its use may be expected to increase as the ready availability of cheap fuel supplies decrease.

Low temperature difference distillation

Low Temperature Difference (LTD) distillation is a means of utilising the low-grade thermal energy content of reject cooling streams from processing plants, diesel engine water jackets, etc. The premise is that this energy will go to waste anyway and if a fresh water supply is required, e.g. shipboard or boiler make-up then a simple flash installation is capable of doing this using the heated effluent as feed.

The design procedures employed for LTD plants vary considerably from those outlined in Chapter 4 for MSF. A conventional MSF plant will operate over a temperature range 127°-32°C (250°-90°F) and heat recovery is of prime importance to keep fuel consumption down. The LTD plant has no such constraint. It commonly may have a heated effluent stream as feed at a temperature of 38°-54°C (100°-130°F). Sea water at *c*. 26.5°C (80°F) may be available as coolant, this gives a maximum temperature differential of 11°C and 27.6°C (20° and 50°F) respectively for heat rejection purposes.

Thus LTD plants commonly operate in the region where conventional processes terminate due to flash chamber sizing, vapour flow problems, etc. The LTD plant is a chamber or series of chambers with provision for feed entry and distribution in a dispersed phase such as a spray or film as hydrostatic head effects dominate at these low temperatures and must be minimised. Cooling is by a simple tube bundle arrangement. The circulation of ambient sea water as coolant may restrict the temperature difference available for operation and the exchanger may require careful design. As the vapour specific volume is correspondingly large at the low saturation temperatures employed efficient demisting arrangements may be required.

Product removal and blowdown can be accomplished by extraction pumps but on land-based installations the plant may be elevated by roughly 10 m and extraction problems eliminated by the use of the resulting hydrostatic head. Such a plant has been designed by the A. Ahlstrom Osakeyhitio Company [15] and

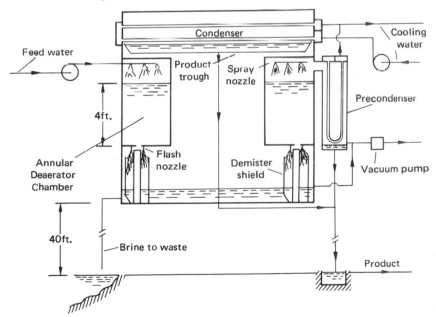

Fig. 6.9. Schematic diagram of low temperature difference plant.

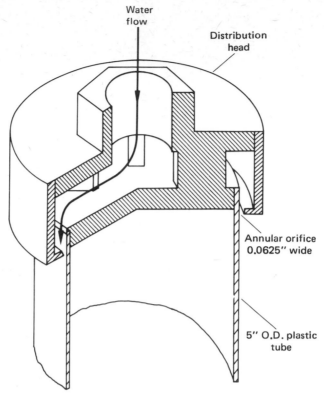

Fig. 6.10. LTD falling film flash nozzle.

produces 100 m³ water per day. The costs claimed for a write-off factor of 15 per cent per year are $0.138/m³ ($0.25 per 1 000 gal) for a 20°C (36°F) temperature differential.

Howe *et al.* [16] have published details of an LTD pilot plant successfully run at the University of California whose details are given in Figs 6.9 and 6.10 which show plant layout and the flow distribution nozzles used respectively.

The simplicity of plants of this type has much to commend it, though as with solar distillation their adoption is a function of the specific situation. The maximum output will be constrained by the feed rate available and permissible flash drop. For a 5.5°C (10°F) flash drop 100 m³ (100 gal) of feed will be required per m³ (gal) of product. The LTD process is another step in the chain of total energy and has little reliance on capital fuel resources given that a heated effluent is being discharged to waste from some other process. Similarly geothermal energy may be employed in LTD installations either as the energy source if steam is vented or as the flashing feed stream.

Controlled flash evaporation (CFE)

Controlled flash evaporation is an attempt to eliminate the hydrostatic head effects which obtain in conventional multi-stage flash distillation. The flashing brine is cascaded down a vertical tower which contains chutes or channels enabling a brine film *c.* 1 mm in thickness to form. The result is controlled equilibration with the vapour substantially in equilibrium with the brine at every point. Although the CFE process is analogous to the LTD process described previously, its use, however, is projected over the conventional temperature range for MSF. The chutes are arranged usually in analogous fashion to the horizontal stages in an MSF plant though the overall stage drop can be 8.3°C (15°F) as opposed to 1.67°-2.78°C (3°-5°F).

The process has been described in detail by Roe and Othmer [17] who see it as a replacement to conventional MSF due to its almost 100 per cent equilibration operation and projected lower construction costs. Simultaneously these authors see it as a potential LTD process operating on heated effluents discharged at 8.3°-14°C (15°-25°F) above ambient. The substantial temperature gradients in tropical seas with surface water up to 40°C above that of the depths are also seen as being suitable for CFE applications – no commercial installations utilise this supposedly free energy source as yet.

Non-metallic heat transfer surfaces

As roughly 40 per cent of the capital costs of most distillation plants is tied up in the heat transfer surface, it is to be expected that many attempts have been made at its elimination. Two routes have been tried, namely the use of an immiscible liquid as a heat transfer medium and non-metallic heat transfer surfaces. Immiscible liquids have foundered technically for many reasons and so far are still at the laboratory stage. Non-metallic surfaces on the other hand are receiving intensive investigation and an assessment of the possible use of glass and thermoplastics in flash plants has been done by Wood [18] and Table 6.3 below (from [18]) summarises the characteristics of typical alloys and non-metallics. Table 6.4 (also from [18]) summarises the assessment for relative costs and characteristics of glass condensers and conventional surface for a (5 m.g.d.) flash plant.

The result of the assessment in Table 6.4 shows a potential 11.8 per cent saving on capital investment compared with the conventional plant. Cheap plastic condensers are so far restricted to a maximum temperature of 71°-75°C (160°-170°F) and the savings are not so marked as those for glass. However, if some of the new varieties of plastics (polyisobutylene, polyphenylene oxide) come down in price the cost picture may change radically.

To date the only reported plant in a commercial size using non-metallic heat exchangers is the 'Kogan-Rose' process, [19] see Fig. 6.11, which uses both direct contact condensation and plastic tubed heat exchangers. The process is in most aspects conventional MSF except that the flashing vapour is cooled by direct contact with a counter current stream of cold product. The heated product is then led to a separate plastic film heat exchanger where it is used to heat the re-cycled brine and sea water make-up stream. By using product water as the coolant stream in the flash chambers costly tubing and water boxes with their attendant problems

Table 6.3. Relative costs and characteristics of plastic and conventional 2 m.g.d. flash plants

Plant		Conventional	Plastic condensers
Flash range	°F	195–105	195–105
Performance ratio	lb/1 000 Btu	7	7
Total number of stages		35	(Conventional 11) (Plastic 24)
Mean overall $\bar{U}$	Btu/ft^2/h °F	600	80
LMTD in recovery stages	°F	9.6	9.6
Total heat transfer surface	ft^2	0.143 x 10^6	(Conventional 0.045 x 10^6) (Plastic 0.750 x 10^6)
Brine flow rate	lb/ft/h	650 000	650 000

Table 6.4. Relative costs and characteristics of glass and conventional 5 m.g.d. flash plants

Plant		Conventional	Glass condensers
Flash range	°F	240–100	200–100
Mean overall $\bar{U}$	Btu/ft^2/h°F.	550	150
LMTD in recovery stages	°F	8.4	6.1
Total heat transfer surface	ft^2	0.556 x 10^6	2.9 x 10^6
Brine flow rate in shell	lb/ft/h	400 000	300 000
No. of decks		2	1
Condenser arrangement		Cross-tube	Through-tube
Wt of shell steelwork	tons	650	775
Approximate shell size	1 x b x h, ft	80 x 45 x 33	110 x 70 x 18
Plant capital cost*	£M	1 144	1 009

*Erected but excluding civil works.
Saving of glass plant over conventional = 11.8% on capital.

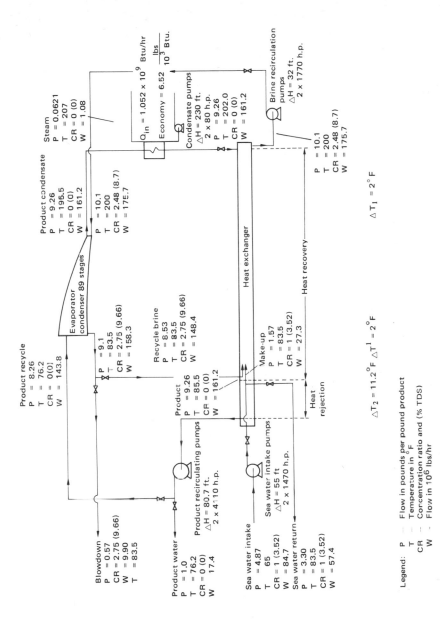

Fig. 6.11. 'Kogan-Rose' 50 m.g.d. (US) process flow sheet.

Legend:
P — Flow in pounds per pound product
T — Temperature in °F
CR — Concentration ratio and (% TDS)
W — Flow in 10^6 lbs/hr

are eliminated. The heat transfer problems are transferred to the design of an external shell and tube heat exchanger. The savings claimed are a 25–30 per cent reduction over conventional desalting plant costs and if the results from a 45 m³/day (10 000 gal/day) pilot plant are successful, the MSF process should have a strong and continuing role in sea-water distillation. As this process has now been running at pilot plant level for over 3 years, it is worthwhile giving the process description and detailed flow sheet for a 50 m.g.d. installation (see 6.11) – the obsession with large-plant capacities seems to be a tendency in most conceptual designs – however, the relevant scale sensitive values may be reduced by a factor of 10 to bring the plant down to present-day capacities.

50 m.g.d. (US) 'Kogan-Rose' process

The plant process is a single-effect multi-stage flash evaporator – direct contact condenser

Sea water at 18.3°C (65°F) is pumped into the plant through a 2.7 m (9 ft) diameter intake pipe from a point located 750 m (2 500 ft) offshore. The sea water is treated with chlorine and screened before entering the sea-water pump.

Approximately two-thirds of the incoming sea water is used for heat rejection. It is returned to the sea at 29°C (83.5°F) after passing through the heat rejection section of a plastic film heat exchanger.

The remaining one-third of the incoming sea water serves as a process make-up stream. It is acidified with sulphuric acid to prevent scale deposition in the plant. It is then partly deaerated in a spray-type deaerator by scrubbing with vapour extracted from the concentrated brine reject stream. Its temperature is raised in the heat rejection section of the heat exchanger to 29°C (83.5°F), the temperature level of the concentrated brine emerging from the last flash stage. The temperature of the sea water make-up stream and of the recirculated brine stream is further raised in flowing through the heat recovery section of the heat exchanger. The two streams emerging from the hot end of the heat exchanger at 95°C (200°F) enter as a combined brine stream into the first flash stage. A small amount of additional deaeration is accomplished in this stage. The hot brine stream flows through the flash evaporator stages which are situated at different elevations, the stage elevation increasing progressively from the hot end to the cold end of the evaporator-condenser. In passing through the successive flash stages the hot brine is progressively concentrated and cooled down due to adiabatic vaporisation. The particular streamlined design of the interstage passages enables the brine stream flowing through each interstage passage to liberate vapour in a quasi-continuous and orderly manner. The mechanical energy liberated during the flash process is spent in lifting the brine level. The brine stream elevation is increased by 3.55 m (11.75 ft) in passing through the flashing stages of the evaporator-condenser. Brine leaves the last stage of the evaporator-condenser at 29°C (83.5°F) and a concentration ratio of 2.75. Of this stream 93.5 per cent flows into the cold end of the heat recovery section of the heat exchanger, while the remaining 6.5 per cent is rejected to the sea after passage through the evaporation section of the deaerator.

The vapour released in the successive flash stage is condensed by direct contact with a stream of desalted water which enters the condensation section of the last (cold end) stage of the evaporator-condenser and flows down through the condensation sections of the successive stages in a direction counter-current to the flashing brine stream. The desalted water stream enters the last stage of the evaporator-condenser at 24.4°C (76.2°F) and leaves the first stage at 90°C (195.8°F). Its elevation is decreased by 10.4 m (34.25 ft).

The distillate emerging from the first stage of the evaporator-condenser flows through a conventional steam heater to the hot end of the heat exchanger. In passing through the heater the temperature of the desalted water stream is raised to 95°C (204°F). In the heat recovery section of the heat exchanger heat is exchanged between the hot distillate stream and the cold brine stream. The temperature of the distillate stream is thereby decreased to 30.5°C (87.5°F). Additional cooling of the distillate is achieved in the heat rejection section of the heat exchanger and it is then pumped from the cold end of the heat exchanger at a temperature of 24.4°C (76.2°F). Of this stream 87.5 per cent is recirculated to the cold end of the evaporator–condenser while the remaining 12.5 per cent is discharged as product.

Summary

Only those processes which have strong applicability and are commercially available or feasible have been described. There are many laboratory variants around, ranging from multiple-effect humidification to the elimination of heat transfer surfaces by the use of an immiscible liquid. Until these are proven commercially a healthy scepticism is usually advisable. Vapour compression and solar distillation are capable of filling a niche in the desalination plant spectrum left untouched by large-scale VTE and MSF plants. The use of solar energy (income), reject thermal effluent or geothermal sources has much to commend it. Capital fuel resources will become scarcer and commensurately more expensive and the solar distillation and LTD processes will gain wider acceptance as the scarcity grows.

References

1. Silver, R. S. 'Fresh water from the sea', *Proc. I. Mech. E.*, Vol. 179, part 1. No. 5. 1964–65.
2. Stormer, R. and Lowes, I. 'A study into nuclear single-purpose MSF plant with vapour recompression', Paper SM–113/52, I.A.E.A. Symp. on Nuclear Desalination, Madrid, Nov. 1968.
3. Wood, F. C. and Herbert, R. 'The characteristics of dual process distillation plant combining vapour compression and multi-stage flash evaporators', Paper SM–113/44, I.A.E.A. Symp. on Nuclear Desalination, Madrid, Nov. 1968.
4. Canberra Department of National Development. A survey of water desalination and the relevance to Australia, 1965.

5. Pilkingtons (Flat Glass Works, St. Helens), Communication, April 1971.
6. Lof, G. 'World distribution of solar radiation', Report No. 21, Engineering Experimental Station, University of Wisconsin, 1966.
7. Porteous, A. 'The design of a prefabricated solar still for the island of Aldabra, 1970', *Desalination* 8, 93–8.
8. Bloemer, J. W. *et al.* Proc. 1st Int. Symp. on Water Desalination. Washington D.C., Oct. 1965, pp. 609–21.
9. Howe, E. D. and Tleimat, B. W. 'Solar distillers for use on coral islands', Paper 99, 2nd European Symp. on Fresh Water from the Sea, Athens, May 1967.
10. Morse, R. N. 'The construction and installation of solar stills in Australia', Paper 101, 2nd European Symp. on Fresh Water from the Sea, Athens, May 1967.
11. Delyannis, A. and Piperoglou, E. *Sol. Energy,* **12**, 1968, 113.
12. Morse, R. N. and Read, W. R. W. *Sol. Energy,* **12**, 1968, 5.
13. Porteous, A. 'The theory, practice and economics of solar distillation', *Chem. Engr.,* **255**, 1971, 406–11.
14. Delyannis, A. and Piperoglou, E. 'Solar distillation developments in Greece', *Sun at Work,* 1st quarter 1967, pp. 14–19.
15. Anonymous. 'Finnish flash distillation', Engineer's Notebook, *Chartered Mechanical Engineer*, March 1973.
16. Howe, E. D. *et al.* 'Distillation schemes using low grade heat energy', Proc. 3rd Int. Sym. on Fresh Water from the Sea, Vol. 1, pp. 691–701, Djbrovnik, 1970.
17. Roe, R. C. and Othmer D. F. 'Controlled flash evaporation on improved multi-flash system', Proc. 3rd Int. Symp. on Fresh Water from the Sea, Vol. 1, pp. 169–90, Djbrovnik, 1970.
18. Wood, F. C. 'An assessment of the possible use of non-metallic heat transfer surfaces in multi-stage flash distillation', Proc. 3rd Int. Symp. on Fresh Water from the Sea, Vol. 1, pp. 363–402. Djbrovnik, 1970.
19. Kogan, A. and Lavie, A. Conceptual design of a 50 m.g.d. MSF, direct-contact condensation desalination plant. Communication from A. Kogan, *The Technion,* Israel, Aug. 1973.

Chapter 7

Part A. Conjunctive use of desalination and conventional water resources*

Preface by author

This material is included to show the application of conjunctive use techniques. The costs used are dated but this does not detract from the validity of the analyses. The Water Research Association specialises in conjunctive studies.

Introduction

To date many of the applications of desalination have been in areas where no significant local water resources exist but where there has been a special incentive to establish or expand a population centre. Common examples are in the Middle East oil communities and in various tourist centres where either the costs of desalination are exceptionally low or the high cost of desalted water for domestic consumption represents a trivial cost to the overall economy.

The growth of these types of application has been startling. However, the contribution of desalination to the world's water resources must remain strictly minimal until desalination can contribute significantly to the increasing demands of existing population centres. For this to take place desalination costs must be reduced still further below the values commonly estimated at present, namely those of operating desalination plants as 'isolated' sources on baseload. Equally it must be recognised that with this expansion in applications the desalination plant will be operating as part of a larger water supply system containing conventional sources as well as the desalination plant.

The particular situation considered here is one where water is supplied from a conventional impounding reservoir and where demands are nearing, or have reached, the maximum sustainable yield of that reservoir. In seeking new supplies

* The majority of this part of the chapter is from a paper by Dr P. A. Mawer and Dr M. J. Burley of the Water Research Association, Medmenham, UK. Desalination 4 (1968) Published by courtesy of Elsevier Publishing Company, Amsterdam.

to meet future increases in demand clearly all possible new sources of water must be investigated. One such possible source may be desalination, and its costs must be compared with those of any alternative supplies. The work reported shows that to make this comparison on the basis of the costs of desalination applicable to baseload operation can be seriously misleading; by the proper design and operation of a desalination plant for use in conjunction with a reservoir, significantly lower costs may be obtained. Considerable economies can be achieved by avoiding operation of the desalination plant during periods when stored water and run-off in the catchment are expected to be sufficient to meet demand. Expectation can be achieved from statistical treatment of historical records. In this way a large part of the operating cost of the desalination plant can be avoided, thereby reducing overall costs to well below those of baseload operation.

Following an outline of the general concept of this type of conjunctive use of desalination, this chapter gives results for two specific examples of how conventional reservoir yields can be increased by the installation of a desalination plant well below the equivalent baseload desalination cost. Both single-purpose and dual-purpose power/water distillation plants are considered; included is the particular case of a dual-purpose system where full electrical generating capacity and no water production is assumed during the 'winter' months (peak electrical demand and/or reduced water demand), and where reduced generating capacity but full water production is available during the 'summer' months (reduced electrical demand and/or peak water demand). Extremely encouraging results are obtained for the two catchments considered, and a generalised method is developed to enable the design of conjunctive desalination systems with any other catchment.

As well as describing the way in which this use of desalination in conjunction with conventional surface water resources can result in greatly reduced costs, the technical implications which this type of low load factor operation has on desalination plant design are also discussed.

The economics of conjunctive use

The conjunctive use of desalination with a conventional source of water represents a special case in the more general field of the conjunctive use of water resources. Work in this field is receiving increasing attention and the techniques appropriate to the design and operation of conjunctive systems are currently undergoing rapid development. In particular the Hydrology Division of the Water Research Association is actively engaged in the development of general programmes by which the conjunctive operation of multiple surface water schemes, or surface and ground water schemes, may be optimised through the use of dynamic programming techniques.

The work reported in this chapter describes a separate but parallel study of the conjunctive use of desalination with natural surface water resources. The work was undertaken as part of the Association's three-year study [1, 2] into the technical and economic assessment of desalination processes and their potential contribution to the future water resources of the United Kingdom.

Conjunctive desalination

In order to describe the advantages of operating a desalination plant in conjunction with a conventional surface water resource it is useful to consider first the normal operation of a reservoir in isolation. Such a reservoir has a 'maximum sustainable yield', which is the yield that can be sustained at all times other than during a severe drought – when some small deficit in the supply must be accepted.* At other times the reservoir may be capable of supplying more than the so-called maximum sustainable yield, as is evidenced by the fact that the reservoir may over-spill during or following periods of particularly high run-off.

Thus the prospect exists of increasing the sustainable yield of a reservoir if a means can be found of supplementing the supply of water during periods of 'drought'. Clearly a desalination plant could fulfill such a role, not being subject to drought in the way that a reservoir is. The potential advantage of such a 'conjunctive' sytem is that the desalination plant need only be operated on relatively rare occasions; the increased total system yield is maintained, nontheless, at all times. In this way the high operating cost of the desalination plant can be largely avoided, and the incremental yield is obtained at a cost considerably below the corresponding base-load desalination cost.

The concept of only operating the desalination plant during 'periods of drought' is clearly an over-simplification. In practice a more precise operating rule must be devised to define when, or under what circumstances the desalination plant must be operated.

A generalised method, for the generation of such operating rules is presented in this chapter.

Before proceeding to a description of the methodology, results of its application to two catchments, of markedly different hydrological character, are described as a means of demonstrating the potential of the concept. It is sufficient at this stage to note that the operating rules developed take the form of sets of monthly 'critical reservoir contents'. The actual contents of the reservoir are inspected at the beginning of each month and if found to exceed the corresponding critical contents, the desalination plant is not operated for the rest of that month.

Conjunctive use of desalination with a short critical period reservoir

The Alwen Reservoir (Denbighshire, N. Wales) provides a convenient example of a short critical period† impoundment in a 'flashy' upland catchment. The reservoir has an average yield of 11 m.g.d., [3] which is equivalent to 79 per cent of the

---

* More generally the yield of a reservoir is a function of the reservoir reliability, increased yields being obtained only at the expense of a reduction in the system's reliability (increased reservoir deficits). The 'maximum sustainable yield' is the yield which can be sustained at an 'acceptable' reliability. In the UK this is usually taken to mean the occurrence of reservoir deficits once or twice in a hundred years.

† This implies that water is stored during the winter for use in the summer of the same year, rather than in subsequent years.

Table 7.1. Costs of water (incremental yields) from the conjunctive use of desalination with the Alwen reservoir

Incremental yield (m.g.d.)	Desalination plant (m.g.d.)	Desalination plant load factor %	Desalination cost* p/1 000 gal	conventional cost† p/1 000 gal	Total conjunctive cost p/1 000 gal	Equivalent base load desalination cost p/1 000 gal
1	1.55	4.26	20	1.17	21.17	44.6
2	3.5	5.6	21.9	1.12	23.02	40.4
2.5	3.8	10.7	22.8	1.08	23.88	39

* For values of $C_1(m, f)$ see †.
† $C_2 = 1.25$ p/1 000 gal – typical of the marginal cost of treatment of an upland water.

Note: All cost values exclude any costs of linking the desalination plant into the conventional system.

long-term average inflow. This is taken as defining the accepted degree of reliability of the system.

In the conjunctive use examples, draft rates have been increased by a uniform daily amount up to a maximum increment of 2.5 m.g.d., corresponding to a maximum total draft of 97 per cent of the long-term average inflow.

Costs of water (of the incremental yield) from the conjunctive systems have been calculated as detailed on p. 125 and are shown in Table 7.1. The operating rules were obtained by using approximate methods, the improved methods having been developed only after obtaining these favourable initial results for the Alwen. The results obtained are, however, entirely feasible but might be subject to some small improvement through the application of the improved operating rule procedure.

The costs shown in Table 7.1 are for single-purpose multi-stage flash distillation with fuel at 1.76 p/10^5 Btu and a fixed charge rate of 12 per cent/annum. Also shown in Table 7.1 are the costs of operating a desalination plant, of capacity equal to the incremental yield, at base-load.

It is seen that conjunctive desalination costs can be as low as 50 per cent or the equivalent baseload cost. Recalculation of the costs at different values of energy cost and fixed charge rate shows the same order of advantage of conjunctive costs over baseload costs.

Conjunctive use of desalination with a long critical period reservoir

The Abberton Reservoir (Essex, S.E. England) is an off-channel (pumped storage) reservoir drawing the bulk of its supply from the River Stour. The Stour contains a large contribution from ground-water sources resulting in long-term persistence of the river flows.

With the existing river abstraction pump capacity (47 m.g.d.), and the particular local requirements of maintaining a certain minimum flow in the river, the yield of the reservoir is 21.3 m.g.d. at an acceptable degree of reliability. This yield represents about 80 per cent of the long-term average inflow which is available when using 47 m.g.d. abstraction pumps.

The important difference between the Abberton and Alwen reservoirs is that the Abberton reservoir has a critical period extending up to three years.

These improved methods have been used to investigate the conjuctive use of multi-stage flash distillation with the Abberton reservoir for the conditions listed in Table 7.2. The cost of taking additional water from the conventional system (C_2), has been taken as 1.66 p/1 000 gal, which includes 0.416 p/1 000 gal for pumping water from the river to the reservoir.

Costs of the incremental yields of the conjunctive systems are shown in Fig. 7.1 for single-purpose plant (fuel price = 1.76 p/therm) and a 12 per cent/annum fixed charge rate. Figure 7.1 also shows the cost of water from a plant of nominal

Table 7.2. Summary of conditions investigated for the Abberton Reservoir

Desalination plant operation	Incremental yield (m.g.d.)	Desalination plant capacity (m.g.d.)		
Unrestricted	1.1	1.22	1.44	1.78
	2.5	2.76	3.26	4.02
	3.6	3.98	4.70	5.80
	4.7	5.22	6.16	7.58
	6.5	7.15	8.45	10.4
	9.0	9.90	11.7	14.4
No winter operation	1.1	1.75	2.08	2.56
	2.5	4.00	4.70	5.80
	3.6	5.80	6.80	8.40
	4.7	7.50	8.90	11.0

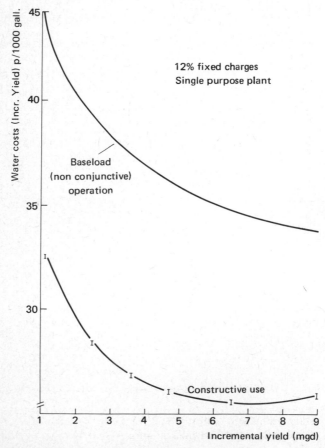

Fig. 7.1. Water cost (incr. yield) pence/1 000 gal.

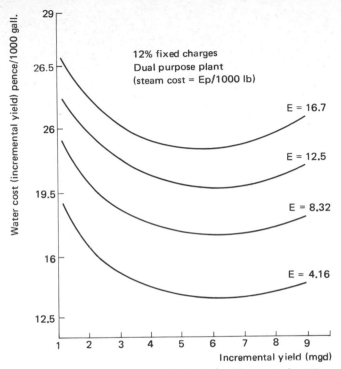

Fig. 7.2. Water cost versus incremental yield (uniform draft, 47 m.g.d. pumps). Unrestricted plant operation.

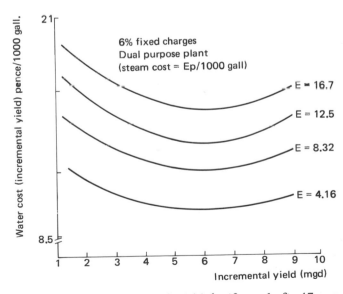

Fig. 7.3. Water cost versus incremental yield (uniform draft, 47 m.g.d. pumps). Unrestricted plant operation.

capacity equal to the incremental yield operated at baseload. Again the use of multi-stage flash distillation is assumed.

From this it is seen that the conjunctive use costs are about 30 per cent lower than the baseload desalination cost.

Figures 7.2 and 7.3 show the conjunctive system costs recalculated for dual-purpose plant operation at 12 per cent and 6 per cent/annum fixed charge rates.

Conjunctive use of desalination under conditions of restricted plant operation

A considerable incentive exists for considering dual-purpose power/water systems in which the desalination plant is not operated during periods of maximum electrical demand in the UK during winter months. Under these conditions, highly favourable rates for the supply of low-pressure steam, to a distillation plant, might be negotiated. Equally, in a power consuming process such as vapour compression

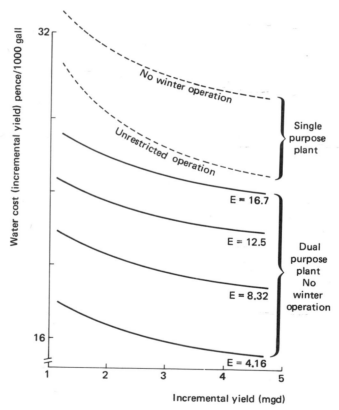

Fig. 7.4. Effect on water cost of restricted plant operation (12 per cent fixed charges, Ep/1 000 lb steam cost).

distillation a more favourable tariff for the supply of electricity might be obtained under these conditions.

Accordingly conjunctive systems have been investigated where operation of the desalination plant is prohibited in months 8–11 (14 October to 3 February) where the year is divided into 13 months (the last month containing an extra day), labelled month 1 to month 13 commencing 1 April. (Demand and reservoir inflows are ignored for 29 February; the available inflow on this date usually exceeds the demand.) The total system water yield is maintained constant, nonetheless, throughout the year. Operating rules for these cases have been generated with the desalination plant capacity set to zero during the prohibited months ($p_i = 0$ for $8 \leq i \leq 11$).

The cases investigated are listed in the lower half of Table 7.2 and costs of the optimised systems, calculated for a 12 per cent/annum fixed charge rate, are shown in Fig. 7.4.

The dual-purpose system with no operation allowed in the winter is seen to be more economic than a single-purpose unrestricted system if pass-out steam is available at less than about 18.75 p/1 000 lb (single-purpose cost calculated at a fuel price of 1.76 p/therm).

Similarly under the conditions considered, restricted dual-purpose operation is cheaper than unrestricted dual-purpose operation if a restricted steam supply can be obtained at 66 per cent or less of the unrestricted supply cost.

Conjunctive use with vapour compression distillation

As stated above, favourable water costs may arise by the conjunctive use of an electricity-consuming desalination process if plant operation is restricted and costs of a vapour compression distillation in conjunction with the Abberton reservoir have been investigated for the conditions shown in the lower part of Table 7.2.

If steam is compressed its temperature will rise and it can then be condensed to generate more steam at the original temperature and pressure. In such a scheme the energy input is in the form of electricity to the compressor.

From the costs of conjunctive use with vapour compression distillation it can be seen that vapour compression distillation cannot compete with single-purpose MSF unless power costs are below about 0.166 p/kWh and 0.25 p/kWh for incremental yields of 1.1 and 4.7 m.g.d. respectively. This is probably due to the lack of economic flexibility available in designing vapour compression units.

The optimum ratio of energy to capital cost is less sensitive to load factor in the case of vapour compression since the compressor cost is a significant proportion of the total plant cost.

The trend of curves in Fig. 7.5 indicates that if a higher incremental yield is taken and a larger plant is used, vapour compression may compete with MSF at higher power costs.

It is seen therefore, that multi-stage flash distillation operated either as a single-purpose plant or as a dual-purpose plant, is more economic in the conjunctive use

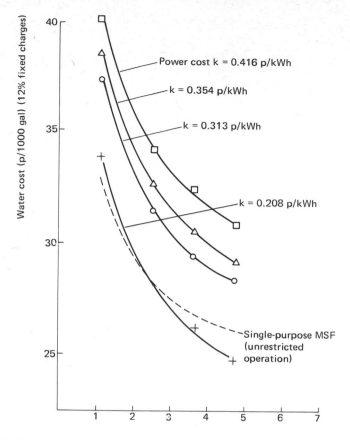

Fig. 7.5. Conjunctive use costs. (Abberton reservoir.) Vapour compression. Summer only operation.

case studies, than vapour compression distillation provided that a steam cost of less than 18.75 p/1 000 lb can be negotiated and as long as power costs exceed 0.25 p/kWh.

The future potential of conjunctive desalination

The conjunctive use of desalination is seen to provide a means of significantly increasing existing reservoir yields at costs well below those of obtaining the same incremental yield from a desalination plant operated at base-load independent of the reservoir. Cost reductions of as much as 50 per cent can be expected in favourable circumstances. Even in the less favourable example of conjunctive use with a long critical period reservoir, savings of as much as 30 per cent have been found.

These cost advantages can be increased if account is taken of plant outages –

which in a baseload application can seriously inflate costs. In a conjunctive system, the full water demand can be met from the reservoir alone during any normal desalination plant outage with no need for the further provision of terminal storage or reserve desalination plant capacity.

It is also to be expected that the conjunctive use of desalination with a short critical period reservoir would be especially well suited to meeting a demand containing a pronounced summer peak.

The further particular advantage of the conjunctive use of desalination is that it makes feasible the design of dual-purpose power/water systems which allow the full generating capacity of the power station to be retained during periods of peak electrical demand – the overall water supply being maintained equally at all times. Clearly this could be of significant economic benefit, particularly in areas where the demands for electricity and water are both subject to peaks but where these peaks are not coincident in time.

The working of further examples is required before the generality of all of the above conclusions can be established. Certainly this is required before extrapolation of the results can be made to arid regions or monsoon areas.

However, the two examples which have been worked go some way to spanning the range of hydrological conditions found in temperate areas such as Europe and the northern coastal regions of the USA. These areas in themselves offer a very large potential for applications. In particular, applications may arise in the more highly developed regions of these countries where existing surface water resources are under considerable pressure from increasing demands.

Developments of this kind could significantly expand the use of desalination beyond the rather special purpose applications which have been common to date.

The impact of low load factor operation upon desalination plant design

As stated above, desalination is in most cases used to provide a year-round supply to areas which have little or no fresh water available or where the cost of such a supply is excessive. However, with the type of use of desalination described here, the plant will not be used at baseload, indeed the load factor may be as low as 10–15 per cent. This does not imply that operation will be at part load for long periods but rather the plant will either be operated at full load or shut-down.

In the conjunctive-use system described above a decision is made as to whether the plant will be operated or shut down for the following 28 days. In the case of the restricted operation study the basic assumption is made that no winter operation will occur.

Corrosion

While in normal operation the saline feed-water is deaerated and hence mild steel can be used providing a modest corrosion allowance is made when designing evaporator shells. If such a plant is shut down for extended periods without special safeguards severe attack of the steelwork will occur. Following shutdown, draining of the vessels, washing with distillate and drying with hot air would do much to alleviate attack and inert gas blanketing would provide further protection. Vertical tube evaporators can be drained more simply than multi-stage flash plants. Sediment which collects in the bottom of flash chambers may prove troublesome and dictate a design which will permit simpler draining.

Product water quality

Distillation plants provide water with up to about 50 p.p.m. TDS and for industrial purposes as little as 1 p.p.m. TDS can be guaranteed. However, the dissolved salts are mainly sodium chloride and the water contains almost no hardness. If such a water is fed directly to a distribution system, it will be aggressive. As a result it will corrode mild steel, cast iron and asbestos cement mains, or will remove organic, inorganic or biological deposits laid down on the mains by the natural fresh water supplied for the greater part of the year. Such attack may result in the premature failure of the mains and severe tastes and odours in the distributed water. The best way to overcome these problems is to blend the distillate with the conventional supply prior to distribution. In some cases blending can be achieved at zero cost but in any case it is unlikely to cost more than 0.416–0.832 p/1 000 gal. As an alternative to blending, the distillate can be artificially hardened by the addition of calcium chloride and sodium carbonate or calcium hydroxide and sodium bicarbonate. [4] Again a small cost is incurred in such post-treatment – further details are included at the end of the chapter.

Plant life-time

Modern sea-water distillation plant dates back a mere ten years and therefore no firm knowledge of plant life exists. However, the general consensus of opinion is that a life of 20 years is reasonable with normal allowances for maintenance including replacement of 50–70 per cent of the non-ferrous tube materials during this period. This results in a fixed charges rate of 12 per cent if interest is at 6.5 per cent (the figure appropriate to the water industry in the United Kingdom).

In the case of conjunctive use an evaporator may only be used for 3 of these 20 years. Provided that adequate shut-down precautions are taken and that scheduled maintenance is carried out, it is quite possible that the life of the plant could be 5 operating years in which case the fixed charges rate would fall to 10 per cent (including maintenance) which corresponds to a capital cost saving of 20 per cent and a water cost reduction of about 10 per cent. A study of off-load corrosion appears justified on this basis.

Computational procedure

Conjunctive use analysis

The computational procedure may be summarised as follows:

1. For the particular catchment (reservoir):
   Calculate values of $_iM_j$ for $1 \leq i \leq n$ and $1 \leq j \leq 13$ where $_iM_j$ is the minimum likely inflow for the period of i consecutive months terminating with month j. The maximum value of i ($= n$) for which values of $_iM_j$ are significant is dependent on the hydrology of the catchment but typically may be anything between 12 and 48.
   Calculate values of $_iM_j'$ for $1 \leq i \leq n$ and $1 \leq j \leq 13$ where $_iM_j'$ is the next-but-minimum likely inflow for the period of i consecutive months terminating with month j.

2. For each combination of draft rate (including seasonal variations) and desalination plant size (including the particular pattern of constraints on plant operation):
   Calculate an operating rule (A) based on $_iM_j$
   Calculate an operating rule (B) based on $_iM_j'$
   Calculate a further four operating rules by simple linear interpolation between the pairs of monthly critical contents of the operating rules A and B.

3. Using each of these six operating rules in turn, simulate the performance of the conjunctive system over an extended time period. Each simulation yields the long-term average desalination plant load factor and the long-term average reliability of the system, measured as the ratio of the actual yield to the design yield. Hence from the set of six simulations performed, the plant load factor corresponding to the desired system reliability can be found by graphical interpolation.

4. Repeat steps 2 and 3 at a new plant size, the draft rate being unchanged.

5. Calculate the costs of the increment yields of each of the systems having the desired reliability (obtained in steps 3 and 4). Incremental yields costs are given by:

$$\text{Cost per 1 000 gal of incremental yield} = C_1(m, f)\frac{mf}{D} + \left(1 - \frac{mf}{D}\right)C_2$$

where

$m =$	nominal daily capacity of the desalination plant (m.g.d.)	
$f =$	desalination plant factor (expressed as a fraction of unity)	
$D =$	incremental yield (m.g.d.)	
$C_1(m, f) =$	cost per 1 000 gal (including capital charges) of water produced in a desalination plant of capacity m operated at a load factor of f. [2]	
$C_2 =$	cost per 1 000 gal of bringing into supply (treatment and/or transmission) any additional water taken from the reservoir.	

6. For the particular incremental yield, determine the minimum water cost, and the corresponding optimum plant size, by interpolation of the values of water cost and plant size obtained in step **5**.

7. Repeat steps **2** to **6** for further values of incremental yield.

8. In the case of an off-channel (pumped-storage) reservoir, steps **1** to **7** are repeated at new values of river abstraction pump capacity (changes in the pump capacity change the values of $_iM_j$ and $_iM_j'$).

9. The reservoir-only system is simulated at a number of draft rates over the same time period used in the simulation in step **3** to yield a number of reservoir reliabilities. The reservoir-only yield corresponding to the desired level of reliability is thereby determined and used as a reference point against which to compare the conjunctive-use system yields – obtained at the same desired level of reliability.

The computational procedure, although simple, is clearly somewhat lengthy. In particular the calculation of the many values of $_iM_j$ and $_iM_j'$, their subsequent use in the construction of an operating rule (as described below) and the simulation of the system over extended time periods, are only suited to computer evaluation.

Generation of operating rules

Let C_j designate the reservoir contents at the beginning of month j. Then for the reservoir not to be in deficit at the end of month j requires:

$$C_j \geq D_j - {}_1M_j$$

where

D_j = the monthly draft rate for month j

$_1M_j$ = the minimum likely inflow for month j

and in general

$_iM_j$ = the minimum likely inflow for the period of i consecutive months terminating with month j.

Any contents below this will certainly result in deficit if the desalination plant is not brought into operation, and if the actual reservoir inflow in month j proves to be the 'minimum likely inflow'. Deficit will be avoided if the plant is switched on when the actual contents fall below $(D_j - {}_1M_j)$ – provided they do not fall below $(D_j - {}_1M_j - P_j)$, where P_j is the monthly output of the desalination plant in month j. Thus the critical contents for month j are given by:

$$C_j = D_j - {}_1M_j \tag{7.1}$$

Considering now the two-month period terminating with month j the condition for the system not to be in deficit at the end of this period gives:

$$C_{j-1} = (D_{j-1} + D_j) - {}_2M_j - P_j \tag{7.2}$$

The term $(-P_j)$ is included since the opportunity exists of re-assessing the criticality of the system at the end of month $(j-1)$ when, if the actual contents are less than $(D_j - {}_1M_j)$, the plant may be switched-on and a quantity of water P_j added to the system.

Similarly, considering the three-month period terminating with month j gives

$$C_{j-2} = (D_{j-2} + D_{j-1} + D_j) - {}_3M_j - (P_{j-1} + P_j) \tag{7.3}$$

These equations may be continued up to:

$$C_{j-k} = \sum_{i=0}^{i=k} D_{j-1} - {}_kM_j - \sum_{i=0}^{i=k-1} P_{j-i} \tag{7.4}$$

These values of C_{j-i} are the reservoir contents required at the beginning of months $(j-i)$ to avoid failure at the end of month j. In principle, values of C_{j-i} must be calculated for all values of j, in order to determine the contents required to avoid reservoir failure at the end of any month in the year. In practice the C_{j-i} for some values of j may be neglected as they do not present operative constraints.

The range of values of i over which C_{j-i} must be calculated is determined by the nature of the particular conjunctive system under consideration. In general C_{j-i} must be calculated for i-values up to the value of the system critical period (expressed in months).

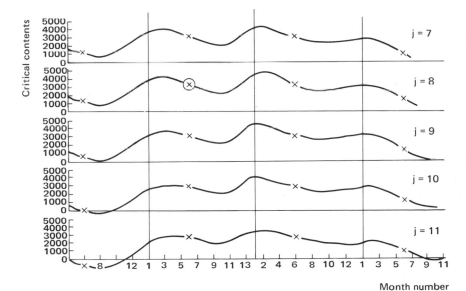

Fig. 7.6. Example of conjunctive use critical contents (C_{j-i}). (Basis is: 24.7 m.g.d. draft, 3.75 m.g.d. plant. Plant ratio = 1.1 and next-min. inflows 47 m.g.d. pumps.)

Results for a particular case are shown in Fig. 7.6 (p. 127). The actual critical contents for month 6, say, for use in the final operating rule is obtained by taking the maximum, shown ($\otimes$) of the 20 values of C_{j-i} shown (x). Similarly maxima are selected from the groups of C_{j-i} corresponding to the other 12 months thereby generating a final operating rule of 13 critical contents.

Closer inspection of Fig. 7.5 shows that, for the example considered, values of C_{j-i} can be neglected for $j < 7$ and $j > 11$; this can be interpreted as meaning that given that the system will not fail in months 7 to 11, then there is no possibility of its failing in other months.

A further improvement in the generation of operating rules could be obtained by taking account of the inter-month correlation between inflows to estimate actual future inflows rather than assume minimum future likely inflows. This has not been considered in this chapter.

Desalination plant size

For any particular draft rate to be sustained, it is clear that the monthly critical contents, must all be below the maximum capacity C_0 of the reservoir. Thus, using equation (7.4) and writing

$$D_{j-i} = D'_{j-i} + \Delta D_{j-i} \tag{7.5}$$

where D'_{j-i} = the reservoir-only draft rate for month $(j-i)$

and ΔD_{j-i} = the incremental draft rate of the conjunctive system gives

$$\sum_{i=0}^{i-k} (D'_{j-i} + \Delta D_{j-i}) - \sum_{i=0}^{i=k-1} P_{j-i} - {}_kM_j = C_0 \tag{7.6}$$

In the reservoir-only case ($\Delta D = 0$ and $P = 0$) the maximum value of the left-hand side of this expression is equal to C_0. Now considering a conjunctive system employing a desalination plant size equal to the incremental yield ($P = \Delta D$), the maximal value of the L.H.S. of equation (7.6) is equal to $C_0 + \Delta D_{j-k}$. The impossibility of having a critical content in excess of the reservoir contents (C_0) shows that this use of a plant size exactly equal to the incremental yield (described as a plant ratio of unity), is not feasible. In principle the plant ratio could be made arbitrarily close to unity by reducing the intervals between decisions on plant operation, thereby reducing the value of ΔD_{j-k}. This, however, does not necessarily lead to an economic optimum, and in any case, can result in an impractically high frequency of plant start-up and shut-down.

As described in step **4** of the computation procedure, all systems have been investigated over a range of plant sizes in order to determine optimum plant ratios. High values of plant ratio lead to low values of critical contents equation (7.4) and in this way the bringing into operation of the plant is delayed, thereby reducing the possibility of switching on the plant only to find that the actual inflows experienced in subsequent months were sufficiently great as to make the plant unnecessary. Against the consequent reduction in operation cost must be set the increase in capital investment resulting from the use of high plant ratios. In general optimum plant ratios have been found to lie in the range 1.1 to 2, although costs are not strongly sensitive to small variations about the optimum.

PART B THE TREATMENT OF DESALINATED WATER FOR DOMESTIC USE*

Introduction

The introduction of desalinated water into piped distribution systems has brought with it problems relating to the protection of pipelines against corrosion as well as the suitability of the product for human consumption.

The technical considerations involved in dealing with these problems are presented in this part of the chapter to amplify the treatment of Mawer and Burley described previously.

Corrosion mechanisms

The water produced by distillation and most other types of desalination plant is very corrosive due to the absence of natural inhibitors and, unless chilled, tastes 'flat' to most palates. It therefore requires suitable treatment when supplied for domestic use through normal supply systems.

Two main types of corrosion can occur when such water is brought into contact with metals, due either to hydrogen evolution or oxygen absorption. The effluent from distillers may contain fair quantities of carbon dioxide due to the decomposition at distillation temperatures of the bicarbonates which are normally present in highly mineralised sources, such as sea water or brackish waters.

The presence of carbon dioxide can depress the pH of the water to a point low enough to enable 'hydrogen evolution' corrosion to occur, provided that the metal stands below hydrogen in the Table of Normal Potentials. This can be likened to the reaction which takes place when base metals are placed in dilute acids, since water and carbon dioxide form the weak carbonic acid thus:

$$H_2O + \quad CO_2 \quad \rightarrow \quad H_2CO_3 \tag{7.7}$$
$$\text{Water} \quad \text{Carbon dioxide} \quad \text{Carbonic acid}$$

The depression of the pH value is caused by the dissociation of carbonic acid into H^+ and CO_3^{--} ions which upset the balance of H^+ and OH^- ions normally present in equal quantities in pure water free from carbon dioxide.

* The assistance rendered by United Filters and Engineering Ltd. in the preparation of this material is gratefully acknowledged.

Metal ions tend to go into solution, leaving electrons behind in the metal. This process would soon come to a halt due to the potential difference that is thereby set up between the liquid and the metal unless the electrons can be removed from the metal. Electron removal is effected by the positive H^+ ions, from the dissociated carbonic acid, which migrate to the negatively charged metal and are discharged electrically as hydrogen gas thus:

$$2H^+ \quad + \quad 2_e \quad \rightarrow \quad H_2$$

Hydrogen ions   Electrons   Hydrogen gas   (7.8)

The point at which the metal ions go into solution and the point at which the electrons emerge may be described as the anode and cathode of an electrolytic cell. Under certain conditions the formation of atomic rather than molecular hydrogen can give rise to polarisation, which means that the electrolytic cell produces sufficient hydrogen to blanket out any further action and corrosion ceases. Dissolved oxygen, however, soon removes this film and corrosion proceeds.

In the absence of carbon dioxide but in the presence of oxygen, electrons are removed from the cathodic parts of the metal by the formation of hydroxyl ions thus:

$$O_2 \quad + 2H_2O + \quad 4_e \quad \rightarrow \quad 4OH^-$$

Oxygen   Water   Electrons   Hydroxyl ions   (7.9)

The hydroxyl ions will unite in solution with the metal ions and produce a metallic hydroxide which, if insoluble, will be deposited at the cathode of the corrosion cell and tend to stifle further corrosion. In the case of iron, ferrous hydroxide and, if sufficient oxygen is present, ferric hydroxide will be formed, both of which are sparingly soluble; the higher oxide being less soluble than the ferrous form. Neither form entirely stifles corrosion partly owing to their porosity, although the ferric form is a better inhibitor than the ferrous form.

Normally carbon dioxide and oxygen are present together and if the carbon dioxide predominates it will convert any ferrous hydroxide formed into soluble ferrous bicarbonate which will remain in solution until such time as sufficient oxygen is available to convert it into soluble ferric hydroxide. In the case of a water pipe this will probably occur some distance from the site of the corrosion cell and therefore exert no inhibiting effect.

Corrosion control

Removal of carbon dioxide and oxygen is the obvious answer to the problem but unfortunately this is only possible in a totally enclosed system free from atmospheric contamination such as a high-pressure boiler feed system which is impracticable for domestic systems.

An alternative answer is to add an inhibitor to the water to form a protective film on the metal. Several inhibitors are available but the choice is narrowed when the water is required for domestic use since the inhibitor must be of proven suitability for human consumption.

An obvious choice is the inhibitor found in most natural waters – calcium bicarbonate.

The presence of sufficient calcium bicarbonate in the water will, under certain conditions, cause the film of corrosion produced – which is laid down in untreated waters in the presence of oxygen – to be modified from the porous hydrated oxide to a 'chalky rust', due to the local precipitation of calcium carbonate. This deposit very effectively inhibits the electrolytic action and corrosion virtually ceases.

An aqueous solution of calcium bicarbonate contains four kinds of carbon dioxide

1. The CO_2 which forms part of $CaCO_3$ (calcium carbonate)
2. The CO_2 needed to convert $CaCO_3$ to $Ca(HCO_3)_2$ (calcium bicarbonate)
3. The CO_2 needed to stabilise the $Ca(HCO_3)_2$
4. Any other CO_2 present which is generally termed the aggressive CO_2 or, as it is in solution, carbonic acid.

Aggressive CO_2 will tend to dissolve any $CaCO_3$ with which it comes into contact and prevent the deposition of a protective film.

The aggressive CO_2 can be removed by passing the water through a bed of marble chips or similar material. This also introduces calcium bicarbonate into the water although the amount introduced does of course depend upon the amount of CO_2 available.

Most modern distillation plants work at low pressure and produce a distillate with a low CO_2 content which is insufficient to dissolve the necessary amount of calcium carbonate to prevent corrosion. In order to obtain sufficient bicarbonate ions to form a reasonable film, CO_2 gas has to be added to the water before it is passed through the beds of marble chips. A very long contact time is necessary to obtain good results and optimum conditions are difficult to maintain,

By-passing some of the raw water round the desalination plant is sometimes practised but although it improves the palatability of the water it does not prevent corrosion because of the adverse ratio of sodium chloride to calcium bicarbonate present in most waters which require desalting.

The conditions under which the desired type of film is laid down are governed by the alkalinity and the CO_2 content of the water, always provided that approximately 50 mg/l of the alkalinity is present as calcium bicarbonate (see Fig. 7.7).

Since the free CO_2 content is almost entirely dependent upon the alkalinity and the pH, we can produce a graph (Fig. 7.8) which shows the ability of the water to produce a protective film as a function of the alkalinity and the pH. Protective films of 'chalky rust' are only formed from bicarbonate waters if the carbon dioxide content is limited to the amount needed to stabilise the bicarbonate so that the smallest rise in pH (caused by production of OH^- ions at the electrolytic cell cathode) is sufficient to render the liquid next to the metal super-saturated with calcium carbonate. It may be thought that any water containing a fair quantity of

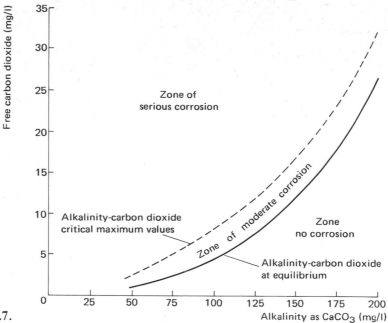

Fig. 7.7.

calcium bicarbonate and having less carbon dioxide than would correspond to equilibrium is non-corrosive but this may not necessarily be true.

A water which deposits calcium carbonate as a precipitate through dissociation of unstable calcium bicarbonate and not as a coating on the metal through cathodic reaction is by no means non-corrosive. The precipitate instead of preventing attack could possibly intensify it by causing differential aeration corrosion cells.

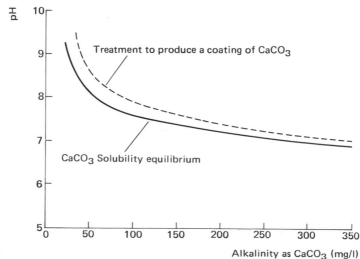

Fig. 7.8.

From the foregoing, it would appear that the best means of preventing corrosion would be to add calcium bicarbonate, remove the CO_2 and adjust the pH and total alkalinity to the appropriate value for laying down a protective film. While doing this, it would be an obvious advantage if some chemicals could be added to the desalted water to make it more palatable and if possible to make it taste like natural water. A commonly adopted system for corrosion control and palatability is shown in Fig. 7.9.

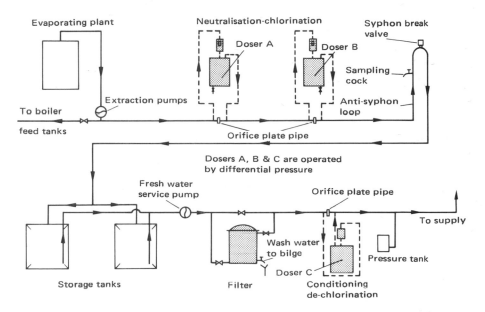

Fig. 7.9.

The calcium bicarbonate is introduced by dosing the water with calcium chloride and sodium bicarbonate which, in solution, are converted to calcium bicarbonate for corrosion control and sodium chloride for palatability thus:

$$CaCl_2 \;+\; 2NaHCO_3 \;\rightarrow\; Ca(HCO_3)2 \;+\; 2NaCl$$

| Calcium chloride | Sodium bicarbonate | Calcium bicarbonate | Sodium chloride | (7.10) |

A little sodium carbonate is also added to adjust the pH to a suitable level and to remove any aggressive carbon dioxide as follows:

$$Na_2CO_3 \;+\; CO_2 \;+\; H_2O \;\rightarrow\; 2NaHCO_3$$

| Sodium carbonate | Carbon dioxide | Water | Sodium bicarbonate | (7.11) |

In the presence of calcium bicarbonate the reaction would proceed thus:

$$Ca(HCO_3)2 + Na_2CO_3 + CO_2 + H_2O \rightarrow$$

Calcium   Sodium   Carbon   Water
bicarbonate   carbonate   dioxide

$$CaCO_3 + 4NaHCO_3$$

Calcium   Sodium   (7.12)
carbonate   bicarbonate

At the same time the water is chlorinated to kill any organisms which could possibly survive the desalting process.

The water is subsequently dechlorinated and can, by the same apparatus, be conditioned with sodium silicate.

The role of sodium silicate as a corrosion inhibitor has been studied by many workers and there is some divergence of opinion regarding the actual mechanism of the formation of the protective film. Studies have shown that sodium silicate, in fairly dilute solutions, forms a mixture of ionic and colloidal silica, the two forms being in equilibrium and instantaneously self-adjusting at low concentrations. Upon formation of solid corrosion products the silica is absorbed or chemi-absorbed by this product to form a thin, self-healing, protective film that does not grow in thickness, consisting of amorphous silica gel together with enmeshed iron oxide and other constituents of the water. The film may be pictured as a two-layer deposit without a sharp dividing line. The lower layer consists of the initial corrosion products. The upper layer is a mixture of an absorption compound of silica and the metal hydroxides enmeshed with silica gel which has extracted small quantities of the iron and calcium compounds, etc., from the water. Only a small amount of silica is necessary to maintain the film and is generally about 8 mg/l expressed as SiO_2. Film formation is faster in hot-water pipes and silica is in fact used extensively as a corrosion inhibitor in hot water as well as cold-water systems. It is unpredictable as an inhibitor in demineralised water but after conditioning with calcium bicarbonate such water is suitable for silicate treatment and this can therefore be added as a valuable second line of defence in case of inadvertent maladjustment of the initial conditioning treatment.

In the unlikely circumstances of a totally closed system 5 mg/l oxygen are required for the treatment to be effective. In normal circumstances the desalinated water will absorb this amount by aeration in storage bonds or storage tanks.

For very large quantities of demineralised water it is more economical to introduce the required calcium bicarbonate by adding lime ($Ca(OH)_2$) to the water, followed by carbon dioxide. The carbon dioxide can be produced economically by burning propane in underwater burners situated in a suitable tank to ensure optimum absorption. This method is usually used in the Gulf States as the CO_2 can be produced very cheaply.

The saving in running costs for large desalination installations soon pays for the higher initial installation costs of this type of plant.

References   135

References

1. Water Research Association, Technical Paper No. 50 Desalination as a supplement to conventional water supply. Part 1—A technical and economic assessment of desalination processes, M. J. Burley and P. A. Mawer, The Association, Medmenham, 1966.
2. Water Research Association, Technical Paper No. 60, Desalination as a supplement to conventional water supply. Part 2 – Desalination developments, conventional water-supply costs and the future of desalination in the United Kingdom, M. J. Burley and P. A. Mawer, The Association, Medmenham.
3. Lewis, W. K. 'Investigation of rainfall, run-off and yield on the Alwen and Brenig catchments', *Proc. Inst. Civ. Engrs.*, 8 (1957) 17–52.
4. Packham, R. F. Aspects of the use of desalted water for public supply. Paper 8, Conference on Desalination as a supplementary water resource, Water Research Association, 1966.
5. US Office of Saline Water R and D Progress Report No. 170. Second annual report brackish water conversion demonstration plant no. 4, Roswell, New Mexico, Washington, D.C., US GPO (1966).

Chapter 8

Combined power and water production

Introduction

The limitations imposed by scaling in distillation processes set 121°C (250°F) as the maximum brine temperature for most plants. The thermal energy in fossil fuel combustion or nuclear reactors is available at a much higher temperature level and as a consequence must be degraded in some fashion for use in a distillation plant. This high-grade energy can be used for power generation in various forms of steam turbine cycles, the steam turbine exhausting or passing out steam at the required temperature to the desalination plant.

The utilisation of turbine exhaust steam can result in greater energy utilisation as the latent heat normally rejected in the condenser cooling water provides the useful heat input to the plant. Typically, in a power-only steam plant, the steam is exhausted under vacuum to around 38°C (100°F) or less if ambient conditions permit and is then condensed in a surface condenser and with heat rejection going to a river or lake or to the atmosphere in a cooling tower. The power-only plant thus rejects 50–60 per cent of the heat input to the plant, other losses average 15 per cent and the resulting overall efficiency is 23–35 per cent.

In a dual-purpose power and water plant the steam can be expanded in the turbine to 121°C (250°F) and sent to the brine heater of the distillation plant.

Because of the relatively low temperature requirements of distillation plants and the low value of energy at these levels the combination of power and water production facilities in one plant may be much more economical than a water-only plant.

Various steam turbine cycles are possible depending on the product ratio, i.e. the ratio of electric power to desalted water.

Combination plants

A combination plant has three major components namely, a heat source, a power generating system, and desalination system. Figure 8.1 shows a basic combination

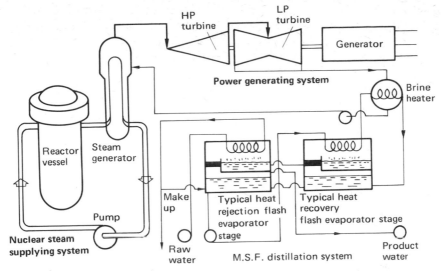

Fig. 8.1. Back-pressure cycle – entire turbine exhaust used in brine heater.

plant employing a nuclear reactor or fossil-fuelled boiler as its heat source, a conventional steam turbine and generator for power production and an MSF plant for water production. In this arrangement all the steam is exhausted from the turbine at *c.* 2×10^5 Pa (29.3 p.s.i.a.) and is used to heat the brine. The back-pressure cycle is the name given to this arrangement and its main advantage is that it has a low product ratio, i.e. produces the least amount of electricity for a given amount of water and is therefore a candidate for adoption in regions where large quantities of water are required but not power. Operational flexibility may be

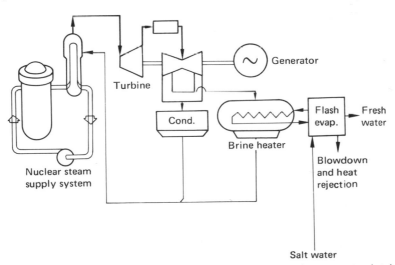

Fig. 8.2. Extraction cycle – brine heater steam is extracted at some point in the expansion.

difficult with this arrangement as the desalting plant cannot be shut down and power only produced unless arrangements are made to condense the exhaust steam.

Figure 8.2 shows an extraction cycle where steam for brine heating is extracted at a suitable point on the turbine the remainder being completely expanded in the turbine and exhausted to a standard condenser. This cycle normally has a high product ratio and is usually specified where small amounts of water relative to power are required. By varying the extraction rate larger amounts of water can be provided when needed.

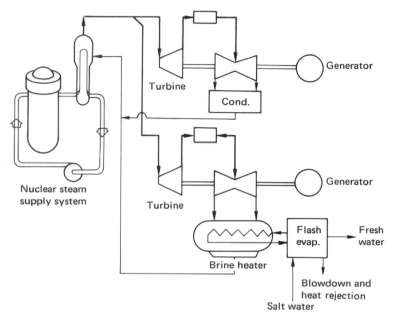

Fig. 8.3. Multishaft cycle – uses parallel condensing and non-condensing turbines.

Figure 8.3 shows a multi-shaft cycle which is basically similar to Fig. 8.1 with a back-pressure cycle operated in parallel with a standard condensing turbine. The water production capability is governed by the back-pressure cycle but power output can be very high. Both the extraction and multi-shaft cycles are capable of flexible operation over a wide range without affecting the distillation plant operation which can be shut down and power only produced.

Nuclear power and desalting plants

There has been and is considerable interest in the use of nuclear reactors for combined power and water supplies. The key to this interest is the difference in the energy cost breakdown for conventional boilers and nuclear reactors. Most conventional boilers have low capital cost and high operating costs except in areas such as the Gulf States where cheap fuel is normally available. Nuclear reactors on

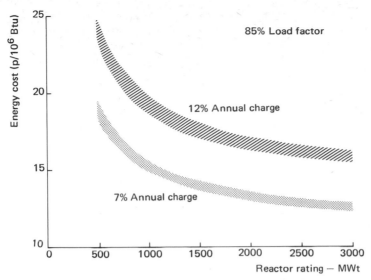

Fig. 8.4. Schematic energy cost versus reactor rating curves for P.W.R. system.

the other hand have high capital and low operating costs. Their marginal operating cost is lower and thus they can normally be used for baseload purposes over a larger period than boilers.

Figure 8.4 shows a diagrammatic energy cost curve for the favoured US pressurised water reactor system. For annual changes of 7 per cent and 12 per cent

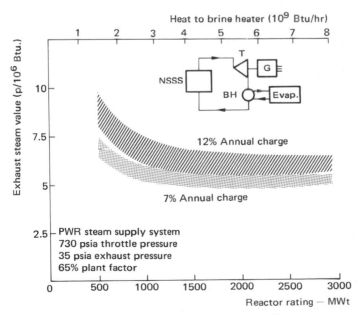

Fig. 8.5. Thermal energy cost from P.W.R. system using full back pressure cycle.

respectively. The principal reason for selecting large capacity plants above 1 500 MW is obvious. Using Fig. 8.4 and adding the cost of a back-pressure turbine generator and subtracting the worth of the electric power to the electric utility (this applies to US conditions) results in the cost of thermal energy at the turbine exhaust. Figure 8.5 shows schematically the result of these calculations. It is seen that scale effects are decreased with a minimum cost band in the 1 500—3 000 MW range.

Another important aspect is that the effect of fixed charge rate is decreased — this is important where dual ownership, as in the US is envisaged with a utility company owning the electricity supply portion of the plant and a municipality owning the water producing portion.

The linking of a turbo-generator and a distillation plant to the same thermal energy source greatly benefits nuclear reactor economics and explains the plethora of feasibility studies in this area. The lower grade energy from a nuclear reactor heat source means a higher steam consumption per kWh generated compared with fossil fuelled plants. As nuclear power plants have lower thermal efficiencies, more heat has to be rejected in the condenser thus a lower product ratio can be obtained. The lower thermal efficiency of nuclear power only plants is eliminated in the case of dual purpose installations.

Cost allocation

There are ten or more different ways of allocating costs in dual purpose installations — Chambers and Wood [1] have considered four basic methods and shown the effect of each on water costs with output for a 1 450 MW advanced gas-cooled reactor heat source with the following ground rules — production cost of electricity assumed as 0.187 p/kWh, capital charge 8.7 per cent/annum and treatment costs .28 p/m^3 (1.25 p/1 000 gal).

The cost allocation bases are given below:

1. *Fixed tariff method.* The reactor is considered to be fixed in size (1 450 MW) and diverting steam and power to the water plant reduces the electrical output,   It is reasonable to maintain the datum electricity cost even at reduced output, and to charge to the water plant all the rest of the expenditure incurred. In Fig. 8.6 curve A represents the change in minimum water cost with increase in output.

2. *Average costs method.* In this case the net electrical output is deemed to be maintained and the reactor size increased as necessary. Production cost of electricity is calculated as reducing with increased reactor size, and the charge to the water plant for steam and power is adjusted to a new scale at each step. Curve B (Fig. 8.6) shows the water cost variation by this method. At very low production of water the figure should agree with that obtained by method A, but as a reactor size and water output increase curve B falls away appreciably below curve A. The reactor size has increased by about 50 per cent at the point where the maximum

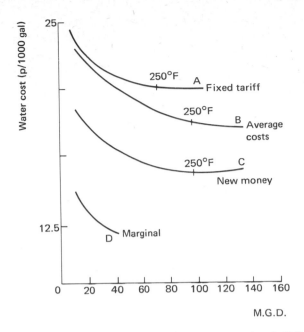

Fig. 8.6. Variation of water costs with MSF plant output using A.G.R. system.

brine temperature is 121°C (250°F). Water output is not 50 per cent greater, however, because of a lower optimum performance ratio.

3. *New money method.* If the cost of electricity is maintained constant, even though the size of the reactor increases, all benefits accrue to water production which carries only the new money or additional cost charges. Curve C indicates how this cost varies with increasing water production, the greatest benefit relative to method B being at low water outputs. As the size of reactor increases the differential between curves B and C decreases because the share of advantage to electricity production in curve B is diminishing in importance and ultimately one would expect these curves to approach and coincide at the point where no power is exported. Thus at very high water/power ratios, if these were achievable, the average cost method which approximates to the US-favoured pro-rating method, almost approximates to the AEA-favoured new money method, while at low water/power ratios the average cost method approximates to the fixed tariff method.

4. *Marginal costs methods.* This method is applicable where advantage can be taken of a small amount of overdesign in reactor capacity and turbine swallowing capacity. No capital charges in these respects are passed on to water production, although increased fuel handling, fuel consumption and related charges are made. Naturally, this leads to the lowest water costs at a given output, but the output is necessarily likely to be very restricted.

The results plotted as variation of water cost against MSF plant capacity are given in Fig. 8.6 and the corresponding plot of optimum performance ratio against water power ratio is shown in Fig. 8.7.

Each curve in Fig. 8.7 comprises two parts corresponding to bled and back-pressure steam conditions at high and low product ratios respectively (or low and high water to power ratios). A full discussion of optimisation of water and power plants based on the nuclear reactor system is given by Chambers and Hitchcock. [2] For a given product ratio, cheap steam leads to a low performance

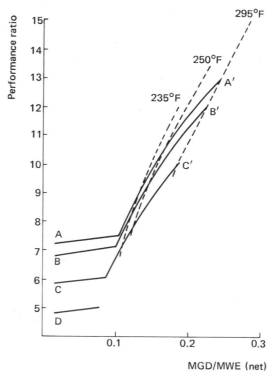

Fig. 8.7. Variation of optimum performance ratio with water/power ratio using A.G.R. system.

ratio desalination plant and expensive steam leads to a high performance ratio. Also as the product ratio becomes smaller and only the expensive higher temperature steam is available the performance ratio must rise accordingly. The product ratio also affects the steam pressure to the brine heater as does the method of costing.

With an appropriate choice of reactor system product, ratios in the range 2.5 to 10 can be obtained which encompasses a wide range of power and water demands. Problems can arise in the flexibility of combined plant operation and there are proposals for the linking of two forms of desalination systems under the umbrella of a dual-purpose plant. These possibilities have been explored by Silver [3] and basically envisage the combinations of power or work consuming desalination processes with low grade thermal energy evaporation. Vapour compression and freezing are two examples of work-consuming processes and thus the flexibility inherent in a mechanical/thermal energy system could allow wide range of product ratios to be obtained from one plant without the need for high performance ratio evaporators or the by-passing of high grade steam from the turbines. Vapour compression combined with MSF has been explored in detail by Wood and Chambers [1] and a persuasive case is made for the adoption of this combination in dual-purpose plants as and when quantities of water (100 m.g.d.) and electric power outputs of 200 MW upwards are required from a dual-purpose plant.

Gas turbine dual-purpose plants

The thermal efficiency of a simple-cycle gas turbine ranges from 20–25 per cent. Various modifications are possible to raise these figures but basically gas turbines have low thermal efficiencies and exhaust gases at c. 538°C (1 000°F) which can be used in a total energy context for steam raising, area heating or distillation. Where there is a requirement for both power and heat the cycle efficiency can rise to 55–70 per cent for a gas turbine with a combined heat recovery steam generator. The addition of supplemental firing for the heat recovery steam generator can increase the efficiency to over 80 per cent.

The economical application of gas turbines for power generation may be very dependent on the exhaust heat recovery for process steam and/or power use.

Let us consider one industrial gas turbine proposal as representative of this class of use, namely a 50 MW gas turbine set for both power generation and as a heat source. The gas turbine is a simple-cycle industrial three-shaft machine with a water-cooled intercooler which is also a heat source between the two compression stages. The nominal performance of the set is 47.8 MW power output and 204×16^6 kJ/h (193×10^6 Btu/h) available thermal energy in the exhaust. The intercooler makes available another 142×10^6 kJ/h (135×10^6 Btu/h). The overall thermal efficiency is 80 per cent with 129°C (265°F) outlet temperature of the exhaust gas.

The plant schematic layout is shown in Fig. 8.8. The brine heater is a two-pass gas/water heat exchanger or a steam generator with finned carbon steel tubes in the gas turbine exhaust stream. This gas turbine MSF layout is projected to provide 41.5 MW saleable power and 26 400 m³/day (5.85 m.g.d.).

Much greater flexibility can be provided if the plant has a vapour compressor driven by the turbine itself and two vertical tube evaporators whose heat requirements are supplied by the vapour compressor as shown in Fig. 8.9. The set can then produce 20 MW of saleable power and 72 000 m³/day (15.8 m.g.d.) of fresh water. Various combinations of vapour compressor MSF and generators can supply different power and water needs.

An industrial community with a high electricity demand will need a high product ratio system. The straight MSF combination would be used and 100 000 persons could be supplied with all power and water requirements, namely, a base supply of 0.4 kW per person and 0.264 m^3/day (58 gal/day). Peaking supplies would also be required close to 1 kW per person – thus an additional 50 MW gas turbine set is needed for baseload variations.

Other combinations of gas turbine and distillation plants have been proposed along similar lines to the above, Carnavos [5] describes a gas turbine heat recovery system with supplemental firing with a claimed overall thermal efficiency of over 80 per cent.

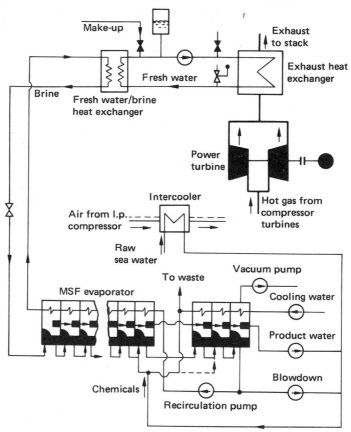

Fig. 8.8. Schematic gas turbine/MSF distillation plant layout.

The combination gas turbine distillation plant system can be attractive in the smaller output range also and some sets are in operation in the 6 MW, 4 546 m^3/day (1 m.g.d.) range. The overall plant efficiency is a function of the exhaust gas exit temperature. This is controlled by the sulphur content of the fuel as cooling below the dew point must not take place or severe corrosion sets in. For natural gas with minimum sulphur content the exhaust gas temperature may be lowered to 129°C (265°F) to give plant efficiencies *c*. 80 per cent. It is interesting to note that the Arabian Gulf State of Qatar is to install 6 x 50 MW gas turbine sets each with an associated 18 000 m^3/day (4 m.g.d.) MSF plant.

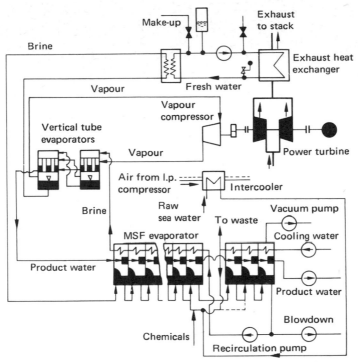

Fig. 8.9. Schematic gas turbine with MSF vapour compressor plant layout.

Total energy and postscript

It is appropriate to end this chapter on dual-purpose plants and the book by an extract from a paper given in 1967 by two men who have done much in the cause of desalination in general and distillation in particular. The extract is from *The Role of Desalination in Water Supplies* by Dr H. Kronberger (deceased) and Prof. R. S. Silver.

> Desalination plants require capital installations and energy. The price of water therefore depends on the annual capital charges and on the cost of energy. A wide range of figures has been quoted for the cost of desalinated water. While it is obvious that the cost of desalination plants must depend on the site, much bigger fluctuations have been introduced through enthusiasm, artificially low interest rate, government subsidies, and commercial risk taking; that is, offering plant at very low cost in order to get into the rapidly expanding business. However, while we are at the moment in the transitional phase, stimulated by the novelty of a recognition that desalination can satisfy urgent needs all over the world, cost estimates will eventually settle down to follow the pattern of any normal industrial development. At the same time, the costs of conventional water have not risen as quickly as originally feared because under the stimulus of a competitive source of water, more careful surveys of exploitation of natural resources are being carried out. Such

surveys nowadays take into account the double function of rivers mainly to supply water, and to act as a transporter of waste; they also take into account the possibility of using the natural river bed for transporting water by pumping additional water into the river and even going from one watershed to the next by subsidiary pumping stations. It might be possible in this way to transport water over much greater distances than previously thought possible. One thing, however, is certain: although the more careful surveys might prevent the cost of natural water rising too steeply in the near future; eventually the cost of natural water will increase; at the moment only cheap supplies are being increasingly utilised and the treatment of effluent and the cleaning up of rivers will cost a good deal of money. Desalination therefore will become economic over increasing areas of the world including those which are not normally classed as arid regions.

We have said before that all desalination processes require energy. This energy can either be in the form of electrical or mechanical energy; or it can be in the form of heat. The process which has been most highly developed and which comprises the majority of the desalination installations in the world is flash distillation; this requires energy primarily in the form of heat. The cheapest form of heat for these plants is the heat which is necessarily rejected from a power station in the course of the production of electricity. A modern coal- or oil-fired power station (and, incidentally, the British Advanced Gas Cooled Reactor) converts 40 per cent of the heat supplied into electricity, and rejects 60 per cent of the energy supplied to waste in the form of low temperature heat. We say that such a power station has a thermal efficiency of 40 per cent, and by present-day standards is quite an achievement. If one wanted to increase the efficiency one would raise the temperature at which heat is supplied to the turbines, or lower the temperature at which heat is rejected. The latter is impracticable since it is determined by the temperature of the surroundings. It is this large amount of waste heat, the inescapable by-product of power production, on which the desalination technologist has focussed his attention. Waste heat at the temperature of the surroundings to which it is rejected has no value. As the rejection temperature is raised, so the heat becomes more valuable for further purposes such as district heating or driving a desalination plant. However, this means that the temperature interval available to the power plant has been reduced and the gain in value of reject heat has been achieved at a sacrifice of power production. Figure 8.10 will make this clear. The left-hand column represents the state of affairs in the modern power station, operating with a steam inlet temperature around 1 000°F (537°C) and rejecting the waste heat to the surroundings at, say 85°F (29°C). The right-hand column shows the state of affairs in a plant where the waste steam is used to drive a desalination plant. Starting again with a steam inlet temperature of 1 000°F (537°C) we now reject the waste heat at 250°F (121°C). In other words, we are passing it to a desalination plant at that temperature. This, of course, can only be done at a sacrifice of electricity production. In fact, we lose about a quarter of the electricity production; the efficiency of electricity production will therefore have dropped from 40 per cent to 30 per cent. On the other hand, we now make *full* use of the waste heat, and now have available 70 per cent of the heat originally supplied to the power station, for the loss of only about one-quarter of the original electrical output. The exact proportioning of costs between electricity production and water production is, of course, quite complicated and is the subject of much discussion between economists.

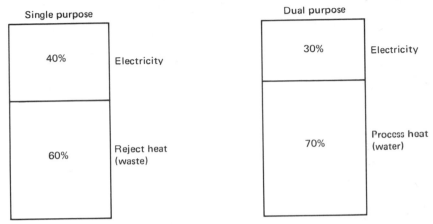

Fig. 8.10. Relative use of effective heat.

... Although we may have appeared cautious in forecasting the role of desalination augmenting water supplies, we are convinced that desalination, will occupy an increasingly important role in supplying water. Desalination, like power production, can be based on coal, gas or oil for fuel; and indeed all existing installations are fuelled in these conventional ways, and such use will extend to still larger sizes of plants. The problem of providing water will eventually move from natural resources of water to resources of fuel. It is in this sense that nuclear desalination will be of particular interest. Abundant supply of low-cost energy and water leads to prosperity – the most solid and practical foundation for peace.

References

1. Chambers, S. and Wood, F. C. 'Economics: Effects of water power ratios, and the role of dual-process dual-purpose plants', *J. British Nuclear Energy Society*, Vol. 7, No. 1, January 1968.
2. Chambers, S. and Hitchcock, A. 'Reactor steam cycles for desalination', Proc. 1st Int. Symp. on Water Desalination, Washington D.C., Oct. 1965.
3. Silver, R. S. Nominated lecture to I.Mech.E., P5/65, November 1964.
4. Lier, N. O. 'Gas turbines prove effective for desalination', *Power,* November 1968.
5. Carnavos, T. C. 'Small/medium dual purpose plants', Int. Conf. on Water for Peace, Washington, May 1967, pp. 198–204.
6. Kronberger, H. and Silver, R. S., 'The role of desalination in water supplies', Paper P/74, Int. Conf. on Water for Peace, Washington, May 1967.

Index

alkaline scale, 27, 28
anhydrite, 35, 36

back pressure cycle, 137
bacterial corrosion, 41
base loss, 103, 104
basin stills, 98–104
bicarbonate, 26, 27, 28
blowdown, 56, 70, 77, 80
boiling point elevation, 20, 21, 22, 61
brine concentration factor, 21, 29, 56, 68, 69
brine heater, 70
bromides, 42

calcium carbonate scale, 27, 28, 29
calcium sulphate scale, 34–37, 91
carbon dioxide, 42, 65, 70, 72, 129
combination power/water plants, 136–147
condensation, 15
conjunctive use, 113–129
controlled flash evaporation, 106
corrosion, 38–48, 123, 124, 129–134
critical period, 115–117
cross tube construction, 66, 67

degassing, 43, 65, 68, 90
demisters, 66, 67
disengagement, 67, 68
dissolved oxygen, 42, 65, 130
dual purpose cost allocation, 140–143

ejector, 70, 90, 91
electrochemical corrosion, 39
electrodialysis, 10
energy reserves, 4
equilibration, 61–65, 71, 72, 107
erosion, 38, 42–47
evaporation, 15
evaporation index, 62–64
extraction cycle, 137

ferric hydrate, 35, 36
first law of thermodynamics, 2

flash chambers, 46, 47
flash nozzle, 108
flash range, 24, 60
flow distribution, 86, 87
fluted tubes, 81–86
foaming, 67, 71
fouling, 17, 18, 25, 71
freezing, 11

galvanic corrosion, 39, 40
gas turbine dual purpose plants, 143–145
gypsum, 35, 36

Hagevap, 32, 33
heat recovery, 18, 66
heat rejection, 18, 65, 68, 69
heat transfer coefficient, 15, 16, 17, 18, 25,
    81, 82, 84, 86
heat transfer surfaces, 15–20, 55, 56, 58, 59,
    60, 61, 68, 80, 81, 82, 89, 93, 94
    metallic, 58, 59, 60, 107
    non-metallic, 107–111
horizontal tube evaporator, HTE, 87–89
hydrological cycle, 4
hydrostatic head, 22, 64, 105, 107

impingement corrosion, 40
incremental yield, 117–123, 125, 126
intergranular corrosion, 40, 41
interstage walls, 66
ion exchange, 31

Kogan Rose process, 109–111

load factor, 123, 124
localised corrosion, 40
logarithmic mean temperature difference, 18,
    19, 20, 59
long tube construction, 66, 67
low temperature difference distillation,
    104–107

magnesium hydroxide scale, 27, 28, 29

maximum sustainable yield, 115
mechanical scale removal, 30, 31
membranes, 11
MSF, 12, 18, 19, 20, 24, 50, 75
multiple effect distillation, 12, 52, 53
multishaft cycle, 138

nuclear reactors, 138–143

orifices, 66, 90

performance ratio, 19, 20, 54
pH control, 33, 34, 38
polyphosphate scale control, 32, 33, 38
precipitation scale control, 38
pressure drop loss, 22, 61
product ratio, 136, 140, 142, 143, 144
product water treatment, 129–134
protective coatings, 43, 46, 47
pumps, 46, 47, 69, 70

recirculated brine, 18, 56, 57, 58, 66–69
recirculation ratio, 24, 58
resources, 1, 3, 4
reverse osmosis, 8, 9, 10, 23

salinity, 8, 26

scale, 25–37, 91, 136
second law of thermodynamics, 2, 23, 24
seeding, 31
solar distillation, 97–104
solar radiation, 98–100
solubility product, 37
splash plate, 66, 67
stages, 54–60, 66, 67, 68
submerged coil evaporator, 22, 23, 52, 53, 54, 55, 56, 77
sulphate reduction, 41, 42

Taprogge, 30, 31, 71
total energy, 145–147
tube plates, 46, 47, 66, 67
tubes, 45, 66, 67, 91

unsuppressed depth, 64

vapour compression distillation, 93–97
venting, 70, 72, 89
VTE, 76–93

water boxes, 43, 66, 67, 71, 90, 91
water quality standards, 6